The Anthropology
of Extinction

The Anthropology of Extinction

ESSAYS ON CULTURE AND SPECIES DEATH

Edited by
Genese Marie Sodikoff

INDIANA UNIVERSITY PRESS
Bloomington and Indianapolis

This book is a publication of

Indiana University Press
601 North Morton Street
Bloomington, Indiana 47404-3797 USA

iupress.indiana.edu

Telephone orders   800-842-6796
Fax orders   812-855-7931

Library of Congress Cataloging-in-Publication Data

The anthropology of extinction : essays on culture and species death /
edited by Genese Marie Sodikoff.
    p. cm.
  Includes bibliographical references and index.
  ISBN 978-0-253-35713-7 (cloth : alk. paper)—ISBN 978-0-253-
22364-7 (pbk. : alk. paper)—ISBN 978-0-253-00545-8 (e-book) 1.
Culture—Philosophy. 2. Anthropology/Philosophy. 3. Extinction
(Biology) 4. Extinction (Psychology) 5. Anthropological linguistics. I.
Sodikoff, Genese Marie, [date]
  GN357.A57 2012
  306.01--dc23
                              2011024026

1 2 3 4 5 17 16 15 14 13 12

A shift in the structure of experience
told the farmer on his Andean plateau
"Your way of life is obsolescent."
—But hasn't it always been so?

"The Displaced of Capital," Anne Winters, 2004

CONTENTS

ACKNOWLEDGMENTS

This book finally came to fruition thanks to everyone's steadfast dedication and to the people who have supported this project along the way. I thank Alex Hinton for having the good instinct to connect me with Fran Mascia-Lees about the possibility of the Anthropology Department of Rutgers-New Brunswick hosting a symposium on extinction that would draw on the strengths of the discipline's subfields. Fran's enthusiastic support and the generosity of the entire Anthropology Department enabled us to have a sustained dialogue about a formative subject of anthropology, an issue of major global significance, and a universal experience of human being. I am indebted to all who participated, including those who do not appear in these pages but who enriched our discussion with their knowledge and insights: John Colarusso, David Hughes, Rob Scott, and Rick Schroeder. I also thank the Rutgers University Research Council and the Center for the Study of Genocide and Human Rights at Rutgers-Newark for additional support. I owe my gratitude to Rebecca Tolen, Molly Mullin, and an anonymous reviewer for their wise and thoughtful suggestions.

And finally I will add, on a very personal note, that during the group's preparation for the symposium, the theme of extinction had become a palpable mood in my home as my mother, Inez Naples Kemptner, who was visiting my family for a last time in the summer of 2008, had grown too weak from cancer to return to Seattle and died in New Jersey that August. She would have been so proud to see this book in print.

The Anthropology
of Extinction

INTRODUCTION
ACCUMULATING ABSENCE:
CULTURAL PRODUCTIONS OF
THE SIXTH EXTINCTION

Genese Marie Sodikoff

In a book published at the cusp of the new millennium entitled *Conversations about the End of Time*, screenwriter Jean-Claude Carrière observes that the future anterior—the tense used to describe an action that will be finished in the future—is fading from everyday speech. He does not comment on the irony that this grammatical form should fall into disuse at this particular time, when projections about earthly life call for such temporal specificity. Scientists have dubbed the current epoch the "sixth mass extinction" because the current rate of species death is more than a hundred times greater than "nature's chronic winnowing" (Angier 2009:3). At some point in the near future, scholars say, 16,928 still extant species *will have vanished* (Zabarenko 2009). At the same time, indigenous languages, vehicles of entire cosmologies, are succumbing at a rate of two per month as their last speakers perish. Of the 6,700 extant languages—already reduced by two-thirds since precolonial times—experts estimate that three thousand *will have gone silent* within thirty years (Miller 2002). Better than any other verb tense, the future anterior captures the jarring imminence of categorical loss. "What are grammatical tenses," asks Carrière, "if not the painstaking attempt of our precise, meticulous minds to envisage all the possible shapes that time can take, all the ways in which we relate to time within the domain of our thoughts and actions?" (Carrière 2001:97).

Thought experiments about what Earth might look like when we too are gone influence decisions in the present (e.g., Weisman 2007). In the midst of a heated politics of global climate change, terrorism, oil spills, war, and the corporate drive to expand into new frontiers of nature, the possibility (and denial) of self-extinction, or at least some dramatic alteration in life as we know it, grips the social imagination. This historical epoch has been named the Anthropocene for the huge impact humankind has made on the Earth's ecosystems (Crutzen and Stoermer 2000). Unlike the first five extinctions (the last being the Cretaceous-Tertiary event that decimated the dinosaurs and enabled the florescence of birds and mammals), the sixth extinction is neither abrupt nor spectacular. No smashing asteroids or giant volcanic eruptions. No global pandemics as yet. Only the slow, cumulative effects of greenhouse gases, rain forest depletion, and a brand of imperialism that extols the virtues of high mass consumption.

What is being done about extinction? What is being thought? Linguists and scientists are undertaking discovery and recovery missions, recording for posterity the last words of indigenous language speakers and the characteristics of rare and "living dead" species, ones destined to die out as a result of habitat degradation (Harrison 2007; Tilman et al. 1994). The United Nations has launched publicity campaigns to stir global interest in the phenomena of language death and biodiversity loss. One may put a positive spin on the process of homogenization, seeing an opportunity for forging closer bonds among societies by ironing out linguistic and cultural difference (see Miller 2002). But the more typical response to extinction is to resist it, or to resist the often flawed strategies to prevent it. In the early 2000s, the United Nations Educational, Scientific and Cultural Organization (UNESCO) created an Atlas of the World's Languages in Danger as part of its effort to preserve "intangible heritage." The UN named 2010 the International Year of Biodiversity in tribute to the problem of species extinctions, which have destabilized the "natural infrastructures" that are so integral to the world economy (Blua 2010). If there is a silver lining in the sixth extinction, it lies in our heightened appreciation of the biological and cultural diversity of Earth's unraveling fabric.

Just as the death of biotic species clears space for emergent creatures, extinction events propel the evolution of cultural productions, including science and technology, politics, history, and art. The prospect of human extinction has animated a doomsday genre of film and fiction, for example. This genre depicts alien invasions and zombie epidemics that annihilate the human species. The life-sucking creatures that fascinate us on the screen and page dramatize and invert the human-nonhuman relationship. From the viewpoint of, say, an Egyptian Barbary sheep (*Ammotragus lervia ornata*), a Guam rail (*Gallirallus owstoni*), or a member of any number of species that have gone extinct in the wild, humans are the monsters to be feared.

The specter of self-extinction in contemporary pop culture might have surprised science fiction writer H. G. Wells, who in an 1894 essay, "The Extinction of Man," opines, "It is part of the excessive egotism of the human animal that the bare idea of its extinction seems incredible to it. 'A world without us!' it says, as a heady young Cephalaspis [an extinct genus of fish] might have said it in the old Silurian sea" (2006:116). Wells follows a long line of European artists and storytellers who found extinct or near-extinct creatures to be rich resources for storytelling. This has been the case for millennia. Historian and folklorist Adrienne Mayor (2001) presents compelling evidence that griffins and giants, phantasmagoric creatures of Greek mythology, likely derive from ancient paleontological speculations on fossils of vanished species.

This volume offers an anthropology attentive to the generative, as well as degenerative, aspects of social and biological extinction. That is, the contributors reflect on the courses of social life that have been enabled or foreclosed by extinction events of the past, and chart the courses that are opening now as the rate and scale of extinction climb. Charles Darwin's (1985) reflections on the laws of species variation in the Galápagos serve as a source of inspiration for this volume. For Darwin, mutations, extinctions, and the formative changes wrought by generations of interspecies interaction were intrinsic to the process of natural selection (see Feeley-Harnik 2007). Natural selection therefore entailed that some species would gradually go extinct as they were supplanted by others (Chernela, this volume). Darwin did not accept evidence of catastrophic mass extinctions, believing instead that "species and groups of species gradually disappear, one after the other, first from one spot, then from another, and finally from the world" (1909:329). This was in spite of the fact that in his lifetime, industrial capitalism in Europe, fueled by imperialism, was changing the planet at an unprecedented pace, and, as scientists were quite aware, species and indigenous cultures were disappearing en masse. Nonetheless, Darwin's view of extinction as intrinsic to evolution offers a fruitful way to think about the coterminous extinctions of biotic species, indigenous cultures, and specific cultural formations insofar as extinction events create voids that direct attention to certain paths forward and are filled by emergent forms of life.

We have also found it fruitful to draw together the strengths of anthropology's four subfields—with an emphasis on cultural anthropology—to address how extinction events have been experienced, recognized, interpreted, and deployed as catalysts for social change, including the social change that results from ecological and genetic restoration projects. The contributors reveal the often unexpected ways in which the destructiveness of extinction to social group cohesion, livelihoods, and ecosystems can simultaneously be productive, insofar as it may yield new thoughts about temporality and existence, inspire creativity, propel technological advancement, and mobilize social movements. The authors explore how people see their own roles, positions, and value vis-à-vis other humans and nonhumans during the mass extinction that is conditioning social and cultural life in real time. The case studies span the globe, including the Caribbean, Native American Canada, Sardinia, and the United States, as well as several iconographic

sites of biotic extinction, including Madagascar, China, Indonesia, and the Galápagos Islands. Some of our key questions are (1) What is the relationship between
the extinction of organic beings and the extinction of cultural formations, such as
languages, ritual practices, and traditional livelihoods? Is it one of analogy, inter-
dependence, or collateral effect? (2) How do projects to retrieve dying forms and
information about past extinction events open up certain paths and close off others? (3) How have extinction events compelled social groups to conceptualize their
place in the world and role in those events?

As the rate of biotic and cultural extinction accelerates, we have been forced
to ponder the meaning of life in its material and immaterial forms, and to imagine ourselves not just as authors of our histories but also as creatures bound by
species-being. The sixth extinction is a species-bound perception of reality. An
interesting fact about the sixth extinction is that it is not defined by a steep reduction of all life on Earth but rather by a reduction in the abundance and diversity of
macroscopic life. Humans, explains biologist Sean Nee (2004), attend to the state
of their food sources, plants and animals, and are repelled by the slimes and oozes
that quiver with invisible microbes. He states, "Our perception of our impact on
the planet as equivalent to a mass extinction simply reflects the evolutionary prism
through which we view life" (Nee 2004:e272).

Due to the scale of human impact on the terrestrial ecosystem, we can no longer perceive ourselves as mere "biological agents" in the world (Oreskes 2007:93).
Although it might take some getting used to, the time has come for us to see ourselves as "geological agents," capable of intervening in and disrupting Earth's geological processes. Dipesh Chakrabarty writes,

> There was no point in human history when humans were not biological agents.
> But we can become geological agents only historically and collectively, that
> is, when we have reached numbers and invented technologies that are on a
> scale large enough to have an impact on the planet itself. To call ourselves geo-
> logical agents is to attribute to us a force on the same scale as that released at
> other times when there has been a mass extinction of species. (Chakrabarty
> 2008:206–207)

In Chakrabarty's view, this transformation in agency collapses the distinction between natural and human history. If the unity of human and natural history must
now be recognized, we might also strive to dissolve the distinctions between the
cultural and the nonhuman, and between material and intangible loss, such as the
memories ensconced in language.

Darwin had already contemplated the process of evolution in a way that dis-
solved the conceptual boundaries between nature and culture in the European

imagination. While cultural group identity is not reducible to language, as is often suggested by campaigns to save endangered languages (see Muehlmann 2008:34), Darwin saw the evolution of language, the prime vehicle of cultural change, as analogous—then homologous—to biological evolution. Elizabeth Grosz argues that for Darwin,

> the development of language is not just *like* evolution, it *is* evolution. Languages, like forms of organic existence, have their own ways of developing over time, their own broad principles, probabilities, and preferences, their own logical or internal force, their impulses to proliferation, which are confronted with the forces of natural selection, that is, with ongoing use in the context of unexpected encounters with hostile or beneficial (linguistic and social) forces. (Grosz 2004:29)

Darwin's model of culture, for which language is the enabling template, contrasts to that of Alfred Kroeber, for whom culture was a "superorganic" entity with an evolutionary trajectory of its own, independent of the mechanism of natural selection. Kroeber (1944) saw the elements of culture in a state of flux, randomly waxing, waning, and being shaped by contact with other groups. The conceptual entwinement of species and culture, and biology and utterance, in Darwin's vision refracts in the contemporary anthropology of human/nonhuman relations.

In recent years, a collective of "multispecies ethnographers" have been examining "the new kinds of relations emerging from nonhierarchical alliances, symbiotic attachments, and the mingling of creative agents" among a multitude of organisms (Kirksey and Helmreich 2010:546). Endeavoring to write an "anthropology of life" in all its entangled complexity (Kohn 2007:4), we tackle the other side—that is, we contemplate the death of forms and the reverberations of categorical loss in social life.

Anthropology and Extinction

The global concern about species extinctions today marks a shift from the nineteenth century, when the extinction problem centered on indigenous peoples succumbing to European expansion. Patrick Brantlinger (2003) argues that a "discourse of extinction" emerged as Europeans waxed nostalgic over the "primitive races" killed by firearms and foreign germs, as well as by the more gradual effects of cultural imperialism, population displacement, and economic and social marginalization. Europeans elegized what they perceived to be living relics of their evolutionary past and regretted the violence done to cultural diversity at the imperial frontiers. Renato Rosaldo reveals the paradox of "imperialist nostalgia":

A person kills somebody, then mourns the victim . . . In more attenuated form, someone deliberately alters a form of life, then regrets that things have not remained as they were prior to the intervention. (Rosaldo 1993:69–72)

Out of the paradox of imperialist destruction and longing for life destroyed, the discipline of anthropology was born.

Ethnography was in part a project to salvage the systems of knowledge and material cultures of the rapidly disappearing indigenous populations. Curators put collections on display in curio cabinets. The artifacts were labeled and their usage described in much the same way natural historians were recording, classifying, and exhibiting specimens of endangered floral and faunal specimens (Urry 1993). Alfred Kroeber, one of North America's disciplinary founders, took a special interest in a man he named Ishi, a last survivor of the Native American Yana people of California. Ishi, meaning "man" in the Yana language, never had the opportunity to go through the Yana naming ceremony because his people had been massacred in 1865. Anthropologists studied Ishi as though he were an archive. He was a living-dead testament to an out-competed and extinguished culture (Kroeber 2002).

By the late twentieth century, anthropologists were championing the rights and sovereignty of indigenous groups. They were intent on preserving indigenous knowledge and practices in situ. Brantlinger remarked in 2003 that the indigenous peoples on Earth totaled perhaps 357 million individuals, most residing in China and India. Even more have been victims of genocidal practices by dominant groups (Brantlinger 2003:190). Rather than merely salvaging the expressive and material culture of vanishing indigenous populations, contemporary ethnography, writes Edward Bruner, has become a "problematic of documenting resistance and telling how tradition and ethnicity are maintained" (1986:140).

In other subfields, anthropologists have approached extinction differently. Biological anthropologists and archaeologists who analyze the fossil and archaeological records offer theoretical reconstructions of the tempo, causes, and effects of faunal (or *Homo* subspecies) extinctions. Biohistorical analyses enable comparative chronologies within the sixth mass extinction. In paleoanthropology, the question of Neanderthal extinction and the relationship between proto-modern humans and modern humans has been the subject of fierce debate. Some scientists have asserted that Neanderthals were out-competed and replaced by modern humans, and others contend that the gene flow between modern humans and their precursors, such as Neanderthal, was continuous (Brauer and Smith 1992). Recent DNA evidence appears to vindicate the continuity model, suggesting that interbreeding between modern humans from Africa and Neanderthals occurred in the Middle East prior to modern humans' expansion into Europe (Than 2010).

Cultural ecologists, more than three decades ago, framed the problem of human extinctions in terms of cultural maladaptation to disasters, comparable to how researchers into climate change undertake analysis of its impact on human societies today (Crate and Nuttall 2009). Critics of cultural ecological analyses faulted functionalist explanations of cultural traits, as well as the assumption of homeostatic processes in the human/environment relationship (see Headland 1997). More recent inquiries into disasters and technological hazards study cultural meanings around extreme loss and change, political and economic factors of disasters, postdisaster social transformations, and assessments of hazard risk.

Susanna M. Hoffman and Anthony Oliver-Smith define disaster as a natural event or process combined with the vulnerability of a social group, "resulting in perceived disruption of the customary relative satisfactions of individual and social needs for physical survival, social order, and meaning." A hazard, in contrast, is defined as a naturally or technologically derived catastrophe that "incorporates the way a society perceives the danger or dangers . . . and the ways it allows the danger to enter its calculation of risk" (Hoffmann and Oliver-Smith 2002:4). Extinction might be considered a slow-onset hazard and disaster. It has uneven velocity and intensity. Rain forests, and the indigenous and nonindigenous groups that depend on their resources, are especially prone to extinction disaster.

Conservation and Indigenous Social Movements

Sometimes anthropological advocacy for indigenous people that is intended to assert their right of cultural self-determination, such as the right to ritually hunt a particular species, clashes with the aims of environmentalists who advocate the right of nonhuman species to exist. Many conservationists support indigenous social movements in principle but may hesitate to endorse the practice of traditional customs that jeopardize already vulnerable species, even if these customs are not the primary cause of species endangerment.

Ecologists define a "keystone species" as one whose influence in an ecosystem is disproportionate to its biomass, helping to determine the types and numbers of other species. The loss of a species may or may not intrude on human awareness. The ripple effects of its absence are determined by its status as what we might call a *cultural* keystone species, one that informs a group's corpus of knowledge, orients symbolic practice, and provides material sustenance. For some cultural groups, particular animals "congeal the complex meanings and struggles about identity and sovereignty" (Field 2008:1). Such animals include, for example, the abalone for Native American populations (Field 2008), the reindeer for the Eveny of Siberia (Vitebsky 2005), and the jaguar for the ancient Maya (Benson 1998). Such species are not only integral to cultural practices and group identities, but they may also have the status of "cultural beings," as in the case of the abalone for Native peoples of California. Les Field explains that this animal retains the status of cultural being even as it "has entered the new century as an increasingly standardized, bio-engineered commodity" (Field 2008:139). Forms of life may also acquire meaningfulness once they are on the brink of extinction, there transforming into entities we consider most valuable and ethically sovereign (Friedland 2006).

A well-publicized case highlighting the politics of extinction with regard to a cultural keystone species is that of the controversy over whaling by the Makah and Nuu-chah-nulth tribes of the Pacific Northwest, who had depended on whaling for centuries until the gray whale population dwindled to near extinction by the 1920s due to commercial hunting. Although the whale had been a focal point of the economy and expressive culture of the Makah and Nuu-chah-nulth peoples for centuries, they voluntarily stopped exercising their government-granted right to hunt for seventy years, until 1994. This is when the gray whale was removed from the list of endangered species. The Makah and Nuu-chah-nulth tribes decided to resume whaling amidst public outcry by environmentalists and animal rights advocates. Their act of reclaiming a political right to hunt also asserted their right to exist as a unique cultural group, and whaling represented an important means of revivifying their cultural heritage (Coté 2010).

As recent cases of Native American whaling have highlighted, the preservation of indigenous lifeways may compel a group to continue to hunt a legally protected species. Even if indigenous subsistence practices are not to blame for the critically endangered status of a particular species, the world economic system that has led to the overexploitation of species also contributes to the marginalization of subsistence-based societies.

On the other hand, indigenous groups may strategically tap into environmentalist discourses to gain global recognition in their own project of cultural survival. For instance, the Penan of Borneo in the 1990s began, in conversations with philanthropic organizations interested in rain forest protection, to make explicit connections between biodiversity protection and the preservation of their indigenous environmental knowledge. By hinging their cultural survival on the conservation effort, emphasizing the link between the preservation of medicinal plants, human well-being, and the preservation of indigenous culture, the Penan and other groups have been able to attract international interest and advocacy (Brosius 1997).

Often such international focus on local ecologies is dreaded by the populations who live there and rely on resource exploitation. Efforts to ward off biological extinction take the form of land enclosure to create nature reserves, the penalization of rule-breakers, and the development of science and technology, such as artificial reproductive technologies, cryobanking, ozone hole repair, cloning, and nanobotics (Heatherington, this volume). Each of these pursuits is rife with moral and ethical quandaries concerning who gets access to natural resources and who loses access, the extent to which scientists should manipulate life for the greater good, and the hierarchy of values ascribed to various species. Is the common good a concept that favors humanity to the exclusion of other species, or geopolitically dominant societies to the exclusion of marginalized ones?

Neil Evernden (1992:6) suggests that a degrading natural environment threatens society's moral order in that ecological deterioration indicates the dissolution of human virtue and a lessened quality of life caused by profligate waste, greed, and carelessness. The recovery and preservation of remnant habitats is a strategy by which institutional actors seek to expiate the wrongs of overexploiting land. Yet the specter of species extinction does not always provoke a feeling of guilt or an act of redemption. Extinction and the ecological changes that result often fortify the survival instinct of social groups, as well as the instinct or desire to preserve cultural identity. This may be accomplished by resisting conservation interventions that recall colonial (and land-degrading) practices.

Overview of the Volume

The volume is divided into four parts, each of which pulls out a thread of connection between biotic and cultural extinction based on analogy, interdependency, or reciprocal effect. Part 1 thematizes the social construction of biotic extinction, emphasizing Western scientific interpretations of extinction events and scientists' responses to and solutions for extinction. The contributors keep in mind the blind spots of scientific theory that often stem from the disavowal of any ideological or cultural bias. Janet Chernela excavates the history of our fascination with species death and its science. She analyzes science as an ideological production that entrenches conceptual boundaries between nature and society. Her account thus reveals the ambivalent acceptance of humanity's geological agency and status as a species linked to the chain of anthropogenic extinctions. On the one hand, science offers the possibility of colonizing space—an escape. On the other, it encourages human beings to consider our species-being as immune from a sixth mass extinction, while cultivating a sense of longing for extinguished life forms and the knowledge they would have bequeathed.

Tracey Heatherington also examines the role of science in producing an extinction discourse and shaping future paths. Drawing on her research in Sardinia, as well as new explorations into the Frozen Ark Project and a global seed bank, she discusses how innovations in genetics, genomics, cryobiology, and assisted reproductive technologies revalorize biological as well as cultural forms. The maintenance of biocultural diversity is reimagined through the lens of genetic management, "genetic rescue," and DNA banking of wild animal species. She contemplates the moral implications and effects on cultural diversity of science's intervention into "the wild." Science contributes to the extinction of the abstract "wild" by reproducing its essence in controlled environments.

Extinction is a process and moment of loss that compels thought about the moral relationships among humans, nonhuman species, and habitats, as well as among social groups with varying degrees of power and autonomy. My own essay, in chapter 3, removes us from the cultural milieu of Western science and examines how science intrudes on the subsistence economy of northeast Madagascar, where rural people grapple with species depletion due to rain forest loss. One finds a convergence of extinction events: Malagasy people are abandoning animal taboos that have long protected certain species from hunting (an instance of cultural extinction), while the populations of endemic fauna are decreasing (an instance of multiple biotic extinctions). I reflect on how conservationists and residents use and transform the meaning of animal taboos and taboo animals, as well as pursuing their respective moral practices, in conservation-heavy zones.

Part 2 concentrates on the ways global biodiversity conservation efforts afford opportunities for marginalized social groups to strategically improve their lot. The global interest in biodiversity protection has attracted international observers to regions with rich and endangered biodiversity. Social groups that have historically lost out to conservation schemes, particularly when land is cordoned off for habitat protection, are finding ways to reformulate their group identity under the international spotlight and to tie their social causes to the global conservation effort and the interest in protecting nonhuman species.

Jill Constantino focuses on the fraught relationship between conservation scientists in the Galápagos Islands, who are trying to preserve a remaining subspecies of tortoise, and Ecuadoran residents of the islands who earn their livelihoods from marine exploitation. Island residents are attempting to stake claims to their cultural practices by tying their history to that of the island's endemic land tortoises and other species. Through the story of Lonesome George, likely the last individual of his subspecies of giant Galápagos tortoises, Constantino explores the society/ species relationship and its politicization. Since a Galápagos National Park worker found George on Pinta Island in 1971, efforts to locate a mate for him have been unsuccessful. Lonesome George's "dynastic" tortoise lineage devalues the status of human residents of the islands who do not participate directly in the species restoration effort. Constantino traces how people are forging a place for themselves in Galápagos history by making kinship with a species on the brink of extinction.

Michael Hathaway describes the emergence of the category of "indigenous people" in China, a concept deemed to be irrelevant to Chinese society until biodiversity conservation efforts opened a political space in which it could take root.

In China, discourses of extinction had relatively little currency until the past two decades, whether describing a disappearing nature or a disappearing sense of cultural difference and authenticity. In southwest China, a convergence of transnational nature conservation NGOs and Chinese social and natural scientists has only recently begun to open ground for an "indigenous space" framed as a source of vital indigenous knowledge and potentially significant cultural rights. Hathaway examines how, in particular conditions, discourses of extinction gain traction through particular sets of academic and activist practices, and how these articulate between international organizations, state officials, and Chinese interlocutors.

In Part 3, Bernard Perley and Paul Garrett address language death, and the languages in question possess radically different values in the discourse of language preservation. While language might mirror the evolutionary process of species transformation, Perley's and Garrett's accounts show how culture can intervene to obstruct the road to language extinction.

Perley, a native speaker of Maliseet and member of the Maliseet First Nation at Tobique, lost his linguistic knowledge at a young age and has embarked on a personal mission of cultural restoration. He examines the process of Maliseet language death and reflects on the collateral cultural extinctions that may accompany it, such as the disappearance of indigenous human rights. The last echoes of the sounds, rhythms, and cadences of the Maliseet language follow colonial erasures of Maliseet place-names and colonial cooptations of Maliseet oral traditions. Perley's intervention is a strategic proposal to recuperate a dying identity. He presents a graphic novel written in Maliseet, which serves to remember Maliseet language, landscape, and sovereignty as everyday practices of indigenous human rights.

Paul Garrett considers the undervalued languages of the world and discusses how the valuation of languages mirrors that of endangered species. The hierarchy of value privileges charismatic megafauna just as it privileges indigenous language. Garrett examines how and why "contact languages," pidgins and creoles, are typically ignored in the discourse of language endangerment and remain on the fringe of the world of languages, including endangered languages. Scholars typically refer to pidgins and creoles as "corruptions" which lack historicity and autonomy, yet contact languages contain a wealth of cultural knowledge and historical experience. They provide a symbolic substrate for historically marginalized people. He advocates for the social, political, moral, and "ecological" value of contact languages. To allow their loss implicitly condones the marginalization and violence experienced by these linguistic communities.

Part 4 situates modern humans in the evolutionary family tree by juxtaposing humans to other primates and hominids. Gregory Forth sleuths the deep past through the ethnological method. Biological anthropologists Laurie Godfrey and Emilienne Rasoazanabary rely on scientific methods to trace anthropogenic causes of ancient lemur extinctions in Madagascar and to study their relevance to contemporary conservation efforts there.

Godfrey and Rasoazanabary rely on paleoecological, archaeological, and paleontological research to construct a detailed chronology for late prehistoric Madagascar. It lends insight into how scientists understand the complex sequence of events that led to one of the most dramatic of recent megafaunal extinction or extirpation crises. We know that most if not all of the large-bodied Quaternary mammals, birds, and reptiles endemic to Madagascar were alive when humans arrived some two thousand years ago. We can, on the basis of evidence from a variety of sources, infer temporal changes in the relative impacts of hunting and habitat degradation; we can also show why large-bodied species are most vulnerable to extirpation and extinction. The past opens a window into the future as we ascertain threats to remaining lemur species, although ethnographic investigation today puts into question the inevitability of their demise in contemporary nature reserves.

Gregory Forth unearths the possible origins of "wildman" tales, stories of zoologically plausible hominid creatures that appear in a wide geographical range of narrative traditions. The chapter offers a cross-cultural account of "wildman" stories. Forth proposes that rather than being pure inventions, these legendary creatures may in fact be correlated with what is known of actual extinctions and recent ecological change, including deforestation and human population expansion. Cryptozoological accounts, generally consigned to the category of folklore, may in fact derive from first-hand witnessing of the last survivors of other hominoid species—ones that are uncomfortably similar to us. His fieldwork findings among people of Flores, Indonesia, suggests that wildman legends might derive from sightings by modern humans of *Homo floresiensis*, a subspecies of hominin (dubbed "the hobbit" in the international press) whose bones were recently discovered on the island.

In his epilogue, Peter Whiteley suggests that efforts to preserve biodiversity and indigenous culture represent a transmutation of the individual's dread of death and desire to conquer it. Social groups find moral purpose as they compete to preserve differentially valued constellations of terrestrial life—elements we distinguish as "nature" and "culture." Landscapes and wild species, with which human beings have a symbiotic and often antagonistic relationship, constantly replenish the wellspring of culture, the essence of social diversity. As Julian Friedland writes, the "massive extinction of biological diversity that humanity is currently bringing about is . . . also the extinction of a part of its own world, that is to say, the extinction of a part of itself" (2006:102). The problem with the eternity of extinction is that when species and languages die, so does the repertoire of genetic and cultural information that has long enabled life to overcome the challenges of survival, as

Celia Lowe argues. Extinctions not only close off the possibility of gathering "scientific evidence of past forms of existence," they also foreclose possible trajectories of evolution (Lowe 2008:108). On the other hand, when social groups perceive and cogitate upon the phenomenon of extinction, some make contingency plans, revising how to proceed and making meaning out of vanishment.

The essays in this collection attest to the high social value of biological and cultural diversity, with endemism and indigeneity forming the positive pole in today's global economy of extinction. Indigeneity and endemic biodiversity are analogous; their intrinsic value intensifies as they are perceived to be endangered. At the same time, cultural and biological systems are interdependent, and extinction processes affect them as an integrated system. Attempts to retrieve cultural forms and biological species play a part in bringing these ideas into currency, conceptually linking diversity to scarcity and making the preservation and celebration of diversity an urgent response to the ever-expansive impoverishment and homogenization of being in the world.

REFERENCES

Angier, Natalie
2009 New Creatures in an Age of Extinctions. New York Times, July 26. http://www.nytimes.com/2009/07/26/weekinreview/26angier.html, accessed April 26, 2011.

Benson, E. P.
1998 The Lord, the Ruler: Jaguar Symbolism in the Americas. *In* Icons of Power: Feline Symbolism in the Americas. N. J. Saunders, ed. Pp. 53–76. London: Routledge.

Blua, Antoine
2010 UN Conference Confronts Dramatic Loss of Biodiversity. Radio Free Europe, October 18. http://www.commondreams.org/headline/2010/10/18-6, accessed October 20, 2010.

Brantlinger, Patrick
2003 Dark Vanishings: Discourse on the Extinction of Primitive Races, 1800–1930. Ithaca, NY: Cornell University Press.

Brauer, Gunter, and Fred H. Smith, eds.
1992 Continuity or Replacement: Controversies in Homo Sapiens Evolution. Rotterdam: A. A. Balkema.

Brosius, Peter
1997 Endangered Forest, Endangered People: Environmentalist Representations of Indigenous Knowledge. Human Ecology 25(1):47–69.

Bruner, Edward M.
    1986 Ethnography as Narrative. *In* The Anthropology of Experience. Victor W. Turner and Edward M. Bruner, eds. Pp. 139–155. Urbana: University of Illinois Press.

Carrière, Jean-Claude
    2001 "Answering the Sphinx." *In* Conversations about the End of Time. Stephen Jay Gould, Umberto Eco, Jean-Claude Carrière, and Jean Delumeau, eds. Pp. 95–170. New York: Fromm International.

Chakrabarty, Dipesh
    2008 The Climate of History: Four Theses. Critical Inquiry 35:197–222.

Coté, Charlotte
    2010 Spirits of Our Whaling Ancestors: Revitalizing Makah and Nuu-chah-nulth Traditions. Seattle: University of Washington Press.

Crate, Susan A., and Mark Nuttall
    2009 Anthropology and Climate Change: From Encounters to Actions. Walnut Creek, CA: Left Coast Press.

Crutzen, Paul J., and Eugene F. Stoermer
    2000 The Anthropocene. Global Change Newsletter 41:17–18.

Darwin, Charles. 1985[1859]
    The Origin of Species by Means of Natural Selection, or the Preservation of Favoured Races in the Struggle for Life. J. W. Burrow, ed. London: Penguin Books.

Evernden, Neil
    1992 The Social Creation of Nature. Baltimore: Johns Hopkins University Press.

Feeley-Harnik, Gillian
    2007 "An Experiment on a Gigantic Scale": Darwin and the Domestication of Pigeons. *In* Where the Wild Things Are Now: Domestication Reconsidered. Rebecca Cassidy and Molly Mullin, eds. Pp. 147–182. New York: Berg.

Field, Les W.
    2008 Abalone Tales: Collaborative Explorations of Sovereignty and Identity in Native California. Durham, NC: Duke University Press.

Friedland, Julian
    2006 Wittgenstein and the Metaphysics of Ethical Value. ethic@ 5(1):91–102.

Grosz, Elizabeth
2004 The Nick of Time: Politics, Evolution, and the Untimely. Durham, NC: Duke University Press.

Harrison, K. David
2007 When Languages Die: The Extinction of the World's Languages and the Erosion of Human Knowledge. New York: Oxford University Press.

Headland, Thomas N.
1997 Revisionism in Ecological Anthropology. Current Anthropology 38(4):605–630.

Hoffmann, Susanna M., and Anthony Oliver-Smith
2002 Catastrophe and Culture: The Anthropology of Disaster. Santa Fe, NM: School of American Research Press.

Kirksey, S. Eben, and Stephan Helmreich
2010 The Emergence of Multispecies Ethnography. Cultural Anthropology 25(4):545–576.

Kohn, Eduardo
2007 How Dogs Dream: Amazonian Natures and the Politics of Transspecies Engagement. American Ethnologist 34(1):3–24.

Kroeber, Alfred L.
1944 Configurations of Culture Growth. Berkeley: University of California Press.

Kroeber, Theodora
2002 Ishi in Two Worlds: A Biography of the Last Wild Indian in North America. Berkeley: University of California Press.

Lowe, Celia
2008 Extinction Is Forever: Temporalities of Science, Nation, and State in Indonesia. In Timely Assets: The Politics of Resources and Their Temporalities. Elizabeth Emma Ferry and Mandana E. Limbert, eds. Pp. 107–128. Santa Fe, NM: School for Advanced Research Press.

Mayor, Adrienne
2001 The First Fossil Hunters: Paleontology in Greek and Roman Times. Princeton: Princeton University Press.

Miller, John J.
2002 How Do You Say "Extinct"? Languages Die; The United Nations Is Upset about This. Wall Street Journal, March 8.

Muehlmann, Shaylih
    2008 "Spread Your Ass Cheeks": And Other Things That Should Not Be Said in
    Indigenous Languages. American Ethnologist 35(1):34–48.

Nee, Sean
    2004 Extinction, Slime, and Bottoms. PLoS Biol 2(8):e272. doi:10.1371/
    journal.pbio.0020272, accessed November 9, 2010.

Oreskes, Naomi
    2007 The Scientific Consensus on Climate Change: How Do We Know We're
    Not Wrong? *In* Climate Change: What It Means for Us, Our Children, and Our
    Grandchildren. Joseph F. C. Dimento and Pamela Doughman, eds. Pp. 66–99.
    Cambridge, MA: MIT Press.

Rosaldo, Renato
    1993 Culture and Truth: The Remaking of Social Analysis. Boston: Beacon
    Press.

Than, Ker
    2010 Neanderthals, Humans Interbred—First Solid DNA Evidence. Na-
    tional Geographic News, May 6. http://news.nationalgeographic.com/
    news/2010/05/100506-science-neanderthals-humans-mated-interbred-dna-
    gene/, accessed November 15, 2010.

Tilman, David, Robert M. May, Clarence L. Lehman, and Martin A. Nowak
    1994 Habitat Destruction and the Extinction Debt. Nature 371:65–66.

Urry, James
    1993 Before Social Anthropology: Essays on the History of British Anthropol-
    ogy. New York: Routledge.

Vitebsky, Piers
    2005 The Reindeer People: Living with Animals and Spirits in Siberia. New
    York: Houghton Mifflin Harcourt.

Weisman, Alan
    2007 The World without Us. New York: Thomas Dunne Books.

Wells, H. G.
    2006[1894] The Extinction of Man. *In* Certain Personal Matters. Pp. 116–120.
    Fairfield, IA: 1st World Library.

Zabarenko, Deborah
    2009 More Than 800 Wildlife Species Now Extinct—Report. Reuters, July 2.
    http://www.alertnet.org/thenews/newsdesk/N01296862.htm, accessed Novem-
    ber 11, 2010.

Part 1. The Social Construction of Biotic Extinction

1. A SPECIES APART

IDEOLOGY, SCIENCE, AND THE END OF LIFE

Janet Chernela

In recent decades science has reached a critical juncture that calls our attention to its fundamental character and the contradictions within it. The crisis was brought about by the observation, by some scientists, that the Earth is facing a massive sixth extinction, one that may have been provoked by human activity. Reaction to this revelation has been complex; it points to some of the ways in which science is influenced by and inextricably integrated into the social fabric.

The degree to which science, as a pursuit of knowledge, is emancipated from the ideological underpinnings of society is an ongoing debate within the social and philosophical disciplines (Althusser 1971; Eagleton 1991; Giddens 1979). Theoretically, science and ideology represent two kinds of knowing, of which the first is open and the second closed. This profound difference has far-reaching implications, suggesting, among other things, that science reaches toward the unknown, whereas ideology continually reproduces itself. From the viewpoint of its proponents, science is an enterprise that not only is open to questions, but is built upon them.

In contrast, ideologies, the idea systems that we rely upon to make sense of the world, are (by definition) closed to challenge. They are intellectual strategies that allow us to organize chaotic phenomena into coherent conceptual schema. Although the reach of any ideology may be sweeping and capable of explaining such matters as, say, "human nature," ideological schema are nevertheless based upon a few generating assumptions. Insofar as they are presented as universal or natural, ideologies discourage, rather than encourage, questioning.

The very act of employing the plural form "ideologies" undermines the totalizing power of any single scheme. The plural is important in this context, because an important source of an ideology's power is its ability to appear as the only option. Anthropological inquiry is, however, based in comparison. It is the comparative viewpoint, present in anthropology since its founding in the nineteenth century, which enables its practitioners to regard all ideologies with the same degree of arbitrariness and to reflect on some of the ideologies that lie at the basis of Western thought and worldviews.

One of the ways in which science may serve society is by contributing to a belief in limitless possibility in which "Man" (First World, Western man, that is) is master of destiny, having attained this position through the domination and accumulation of knowledge. A fundamental tenet of this ideology is the definitive separation of humans from all other creatures, and the earned status of humans as inheritors of the Earth's bounty. From this perspective, therefore, science may constitute part of the larger project of technological progress and Western hegemony that drives global economies of growth and celebrates human achievement.

Over the past two centuries, scientific accomplishment has provided a principal resource for optimism, confidence, and the celebration of Western achievement. That science has brought a sense of life, vitality, and possibility to the public has been apparent for some time. Its role in the forward motion of material accumulation and resource exploitation has been fundamental to its momentum. The science of extinctions, as findings and as sets of ideas, would thus appear to be an exception to the pattern.

A dramatic reduction in global biological diversity, brought about in part through activities deemed to be the fruit of advances in scientific knowledge, calls into question our assumptions about the scientific enterprise and the uses to which it is put. In this chapter, I briefly review the growing awareness of extinctions in order to consider what it tells us about the role of science in society.

Icons of Extinction: *Objets Morts*

Visitors to the *Wunderkammern* and *Cabinets du Roi* (curiosity cabinets) of the eighteenth and nineteenth centuries delighted in *objets morts*—the dried, stuffed, and bottled remains that evidenced a mysterious past (Olalquiaga 2006). Contained among the collections of *naturalia* were unidentified, mystifying objects from excavations. As citizens of a newly found progress, Europeans were fascinated with things passed, or *morts*. The greater the contrast with the present, the more entrancing the object. By looking back at a past populated by beings of grotesque difference, humans could place themselves at the apical meristem—the growing tip—of the future.

Figure 1.1. Ole Worm's cabinet of curiosities. From *Museum Wormianum* (1655), in the possession of the Smithsonian Institution Libraries.

The discovery of the skeletal remains of the American mastodon in 1705 served as evidence of the existence and disappearance of an intriguing species. Exhibits of mastodons captivated Europe during the eighteenth century. In the American colonies the discoveries of mastodons coincided with a growing independence movement and a new national identity. In 1801 Thomas Jefferson, an avid collector of fossils and a student of extinctions, joined Charles Willson Peale in an expedition to exhume mastodon bones in upstate New York. The achievement was regarded as one of the important events in the history of American science and Peale mounted the gargantuan skeleton in his Philadelphia Museum, one of the first natural history museums in the United States. The specimen, displayed along with a murderer's finger and an eighty-pound turnip, was considered the museum's first successful attraction. The sensation was auctioned in 1849 and sold to bidder P. T. Barnum.

Peale wrote that the mastodon exhibit was a spectacular example of what he called democratic access to knowledge within a private institution (Barney 2006). Admission tickets to Peale's Philadelphia Museum bore the words "The Birds and Beasts will teach thee!" (Yanni 1999:28). Years before the publication of *On the Origin of Species,* fossils such as these served to inspire conjecture about species change. Knowledge inhabited the very materials that served to convince onlookers of earlier species' existence, and by inference, their demise. By the mid-nineteenth century a European subculture of gentleman fossil hunters had emerged. It was one of these, William Parker Foulke, who found the first nearly complete dinosaur skeleton while visiting Haddonfield, New Jersey, in 1858. Foulke's findings were stunning. As he noted, the bones suggested an animal larger than an elephant that combined the structural features of lizards and birds. There was no context in which to comprehend a relationship between genera as apparently distant as these. Dinosaur fossils had been known in England since 1677, yet they went relatively unnoticed until 1824, when the accumulation of specimens and concomitant interest in them was sufficient to derive patterns and draw inferences. The evolutionary affinity between dinosaurs and birds was not understood until the end of the twentieth century, when it was hailed as a new finding, a contribution to the cumulative, forward movement of scientific knowledge.

Another icon of extinction, the flightless dodo (*Raphus cucullatus*), was discovered in Mauritius in 1581 but went extinct soon after with little public outcry. The demise of the dodo is directly attributable to the arrival of Europeans who overhunted the birds, destroyed their forested habitats, and introduced dogs, pigs, cats, rats, and macaques onto the island, where they ravaged nesting sites. The extinction of the dodo in the seventeenth century received little attention until the 1859 publication of Darwin's *On the Origin of Species*. In the surge of interest it prompted in species change and disappearances, new caches of dodo bones were unearthed and described to an avid public. In 1865 Lewis Carroll gave the dodo further prominence in *Alice's Adventures in Wonderland*. Since then the dodo has come to stand for extinction. Today a number of environmental organizations use the image of the dodo to promote the protection of endangered species.[1]

Charles Darwin, like others of his time, was intrigued by the fossils of extinct animals and plants. His observations of fossils and their comparisons to living species during his explorations on the H.M.S. Beagle contributed to the development of his theory that all life on earth evolved from a few common ancestors. Darwin's evolutionary theory, which he published as *On the Origin of Species* in 1859, identified extinction as one of three principles underlying species change over time. Natural selection entails extinction; it brings about the survival of some species or varieties, but inevitably causes the extinction of others: "as new species in the course of time are formed through natural selection, others will become rarer and rarer, and finally extinct" (2010:48). He later continues, "Within the same large group, the later and more highly perfected sub-groups . . . will constantly tend to supplant and destroy the earlier and less improved sub-groups . . . [W]e already see how it [natural selection] entails extinction; and how largely extinction has acted in the world's history, [as] geology plainly declares" (2010:58).

In order to consider "whether species have been created at one or more points on the earth's surface," Darwin observed the distributions" of species on continents and islands and suggested that the relationships between island species and those of the nearest continent could best be explained by processes of migration and natural selection.

> A volcanic island, for instance, upheaved and formed at the distance of a few hundreds of miles from a continent, would probably receive from it in the course of time a few colonists, and their descendants, though modified, would still be plainly related by inheritance to the inhabitants of the continent. (Darwin 1911:351)

Using the model of the continent/island relationship, Darwin was able to draw generalities about the two principal axioms on which evolution is based: diversification and extinction.

Darwin wrote candidly about his personal challenge in assigning such prominence to the role of destruction in his theory.

> Nothing is easier than to admit in words the truth of the universal struggle for life, or more difficult—at least I have found it so—than constantly to bear this conclusion in mind. Yet unless it be thoroughly engrained in the mind, the whole economy of nature, with every fact on distribution, rarity, abundance, extinction, and variation, will be dimly seen or quite misunderstood. We behold the face of nature bright with gladness, we often see superabundance of food; we do not see or we forget, that the birds which are idly singing round us mostly live on insects or seeds, and are thus constantly destroying life; or we forget how largely these songsters, or their eggs, or their nestlings, are destroyed by birds and beasts of prey; we do not always bear in mind, that, though food may be now superabundant, it is not so at all seasons of each recurring year. (2010:26–27)

The publication of *On the Origin of Species* in 1859 is a recognized watershed in biological science. Perhaps the greatest threat to Western ideology was not the common origin of all beings, as is assumed, but rather the possibility of a common ending: that all beings, humans among them, were subjected to the same forces and vulnerabilities. Insofar as it challenged orthodoxies, the appearance of *Origin* may be considered a "natural experiment" in the articulation and confrontation of different kinds of knowledge, open and closed. With its publication, social subjects were faced with their own conceptualized models of life on Earth and the place of humans within them.

The publication of *On the Origin of Species* represents a defining moment in science, at which it was confronted with the limits of its own autonomy. The processes and principles put forth in *Origin* could be extended to humans, but Darwin deliberately chose not to include that species in his descriptions and observations. At the time of his book's publication, the place of humankind in the context of living creatures was a matter of heated debate. Darwin's observations, if extended to human beings, would have been a direct challenge to the prevailing ideology. To avoid offense or sanction, Darwin, rather than entering the debate, left a hiatus in the text where a mention of humans might have been (although he was to correct this in subsequent publications). The deliberate omission underscored the importance of the absent message; the silence made by it was louder than any utterance might have been.

Preservation

The last half of the nineteenth century was a period of heightened interest in, concern with, and debate on the processes that drive species change. At the same time transportation innovations, industrial expansion, and empire building increased European contact with the periphery. The economic expansion was accompanied by a rush to collect knowledge about species never before known to Europeans, as well as concerns about species decline and efforts to protect wilderness areas. In 1866 the British Colony of New South Wales in Australia created Blue Mountains National Park, the first modern national park (Phillips 2004). This was followed by a series of parks, including Yellowstone National Park in the United States in 1872; Royal National Park in Australia in 1879; and Bow Valley, now Banff National Park, in Canada in 1885.

The following decade, between 1892 and 1902, witnessed the founding of the world's first environmental organizations. The Sierra Club, founded in 1892 by the Scotsman John Muir, then living in California, was created to preserve wilderness areas as places of repose and beauty. In 1893 the British founded the Royal Society for the Protection of Birds and in 1884 the National Trust for Places of Historic Interest and Natural Beauty (Epstein 2006:37). In 1902 the European nations signed the first international instrument to protect animals, the Convention to Protect Bird Species Useful in Agriculture (Epstein 2006:37).

Often cited as the earliest and longest-lived international animal protection association, the Society for the Preservation of the Wild Fauna of the Empire was founded in 1903. The goal of its founders, British colonial officers and naturalists working in Africa, was to minimize the impact of local hunting on the large game populations of East and South Africa. In the service of recreational hunters and in opposition to subsistence hunters, the society established legislation that limited hunting to designated park lands. These activities laid the foundational framework for the creation of large national parks, such as the Serengeti game reserve established in 1921, which would become the well-known Serengeti National Park of Tanzania. Today the organization is known as the Fauna and Flora Preservation Society, with members in more than eighty nations. These early international conservation advocacy organizations established a precedent and groundwork for the intergovernmental agencies to come, including the International Union for the Protection of Nature (now the International Union for Conservation of Nature and Natural Resources, IUCN), the World Wildlife Fund (now the World Wide Fund for Nature, WWF), and the Convention on the International Trade in Endangered Species of Wild Flora and Fauna (CITES).

Europeans were familiar with loss of fauna through overexploitation. Large fauna had gone extinct from overhunting in Britain when the industrial revolution brought towns, roads, and railroads within range of game animal habitats. When the deer population went into decline, foxes replaced deer as the recreational prey. By the late nineteenth century the popularity of fox hunting had begun to deplete the fox population as well, forcing the British to import foxes from the continent.

Quantifying Extinctions

As frequencies of species extinctions rose with the expansion of industrial economies throughout the world, attention shifted from individual extinctions to rates of extinction. A new environmentalism was marked by the exploration of survival and variation. The paradigm made these phenomena measurable, producing new indices, including rates of species loss and recovery, as well as of intra-species and extra-species variation. Among the innovations of the latter decades of the twentieth century is the concept of biodiversity, a contraction of "biological diversity," which assigns value to variation per se and equal worth to all living things.

William Hornaday, a zoologist and conservation activist who was greatly concerned with vanishing species, is credited with compiling the first list of extinct species. As the director of the New York Zoological Park in the Bronx, Hornaday revolutionized the practices of zoo exhibition by displaying fauna in their natural settings. In one such exhibit in 1906, Hornaday placed Ota Benga, a native of the Congo Ituri forest, on display in the monkey house, where, before crowds of spectators, Ota Benga aimed a bow and arrow and tussled with an orangutan (Bradford and Blume 1992). In order to draw public attention to the increasing endangerment of wildlife, in 1913 Hornaday published *Our Vanishing Wild Life*. The following year he published *Wild Life Conservation in Theory and Practice,* in which he listed ten species that had become totally extinct between 1840 and 1910. The book, aimed at educating the public to save wildlife from extinctions, was read widely.

Concern about rising extinction rates was one of the driving forces behind the 1949 creation of the International Union for Conservation of Nature and Natural Resources (IUCN), whose Species Survival Commission (SSC) was charged with monitoring the threatened status of mammal and bird species. The charge was expanded in 1963 when the eighth general assembly of the IUCN, held in Nairobi, established the Red Data Book, an inventory of threatened species. SSC scientists would head the project, attempting to address gaps in the knowledge of threatened species throughout the world (Davis 2007:169–170). The creation of the Red Data Book, later to be known as the Red List of Threatened Species, constituted a landmark in measuring and reporting global extinction rates.

The Red List of Threatened Species, which assesses and ranks the vulnerability of plant and animal species, was created to reduce species extinctions and convey the urgency of conservation issues to international civil society and policy makers. Initially, SSC assessments were based on data provided by BirdLife International, the Institute of Zoology (the research division of the Zoological Society of London), the World Conservation Monitoring Centre, and specialist groups within the SSC itself. Today the work of the SSC relies on some seven thousand species experts and partner institutions in almost every country in the world.[2] Collectively, this extensive scientific network cooperates to produce a cumulative corpus to document the changing conservation status of species. The system is now considered a standard conservation instrument to assess extinction risk for species in taxonomic groups worldwide. There are nine categories in the IUCN Red List system: Extinct, Extinct in the Wild, Critically Endangered, Endangered, Vulnerable, Near Threatened, Least Concern, Data Deficient, and Not Evaluated.

The Determinants of Extinctions

Noting and listing extinctions was one thing. Understanding why they occurred was another. While the fossil record showed ample evidence of extinct species, some researchers were concerned with looking into the determinants of extinctions to derive models that would explain past extinctions and predict future ones. One group of researchers investigating the conditions that drive local extinctions were adherents of an emerging field known as island biogeography. These researchers, who include Robert MacArthur, E. O. Wilson, Daniel Simberloff, and Jared Diamond, laid the basis for the new field of conservation science.

Building on Darwin's insights and observations of oceanic islands, these authors generated a number of hypotheses and models to predict rates of extinction and species colonization in island habitats. The field had its beginnings in the 1963 collaboration of Robert MacArthur and E. O. Wilson in developing a model that would predict rates of extinction and colonization based on the size of an island and its distance from a colonization source (MacArthur and Wilson 1963). In 1967 MacArthur and Wilson published their observations and hypotheses in a full-length book, *The Theory of Island Biogeography*. In it they proposed that an island's size and distance from a source community are the two principal factors determining its species numbers, diversity (variation), and rates of recolonization. These factors, they argued, determined the rates of extinction and immigration within any circumscribed habitat. Their model of species-area relationships generated a number of subsequent studies intended to support, refute, or refine it.

One of the earliest of these was a 1969 study conducted by E. O. Wilson and Daniel Simberloff in which they fumigated mangrove forests to mimic conditions of extinction, then monitored the recolonization of the mangroves by insect and arthropod populations (Simberloff and Wilson 1969; Wilson and Simberloff 1969; Simberloff and Wilson 1970). In 1972 the young professor Jared Diamond found a natural experiment in the massive defaunation event caused by the 1883 explosion of the Krakatoa volcano in the Pacific. Diamond was able to measure plant and animal regeneration on 15 of the islands affected by the explosion (Diamond 1974). These and other observations contributed to an understanding of changes in species numbers and distributions, including rates and patterns of colonization and reproduction, in response to differing factors. In addition to scholarly articles, a number of popular works have emerged from the interest in island biogeography. Among them is David Quammen's *The Song of the Dodo: Island Biogeography in an Age of Extinctions* (1997).

The implications for conservation were apparent. In 1975 Jared Diamond linked the principles of island biogeography to conservation policy and park design in a foundational paper, "The Island Dilemma: Lessons of Modern Biogeographic Studies for the Design of Natural Reserves." In it he addressed the shrinkage of natural habitats and biodiversity loss resulting from human-induced landscape alteration. Diamond was concerned with calculating the rate at which human activity might transform landscapes, rendering them isolates. Using an area-species formula based in the MacArthur/Wilson model (which was later criticized and refined), Diamond drew a number of inferences to be used in the planning of natural reserves. He also called attention to the speed with which formerly continuous natural habitats and distributional ranges of human-intolerant species were being fragmented into disjointed parcels (Diamond 1975:130). These insights, and Diamond's suggestions for remediation, laid the foundation for growing interest and research in corridor ecology, a field that has gained considerable attention in natural reserve design planning.

Diamond applied the island model both literally and as a metaphor. Marine islands are expanses of firmament delimited by surrounding water, but many other situations possess the same distributional significance for biota. We can therefore extend the term "island" metaphorically to any habitat that is surrounded by conditions that are inhospitable to species residing in that habitat. By this extension, many habitats may be regarded as islands, because their limits constitute a barrier to the distributions of species. Accordingly, for a parasitic species the body of the host organism is a distributional "island" surrounded by a "sea" of inhospitable conditions; for an aquatic species a body of water is a distributional island surrounded by a "sea of land"; for a crop the cleared field is a distributional island in a forest "sea" of aggressive competitor species (Diamond 1975:129–130).

The implications extend, too, to population dynamics in any biological community surrounded by inhospitable conditions resulting from human alteration. In this way the Earth can be seen as an isolated complex of ecosystems surrounded by interplanetary space. In the 1970s James Lovelock proposed that the Earth—which he called Gaia, after the Greek deity—could be regarded as a single complex ecosystem. Although this suggestion aroused some skepticism, the term "biosphere," which had been introduced first in 1875 by Eduard Suess, was increasingly coming into use. Today it is no longer novel to regard the Earth as a circumscribed, isolated, and fragile habitat. Most recognize that the Earth is a "small planet" (Lappé 1991). In the sense used by the island biogeographers, the Earth is an island.

Sounding the Alarm

Awareness of soaring extinction rates resulting from human activity contributed to the expansion of environmental advocacy institutions, nongovernmental organizations (NGOs), and programs in the 1960s and 1970s. The

World Wildlife Fund, created by supporters of IUCN in 1961, was founded to raise private funds toward protecting endangered environments and threatened species. In the beginning of the 1970s WWF teamed up with the then-new Natural Resources Defense Council (NRDC) in a large project entitled Threatened Species and Genetic Resources. WWF had estimated an extinction rate of one species per year, as compared to one every ten years between 1600 and 1950 (Kreisler 1999). But a young consultant to NRDC thought that estimate was too low. The consultant, Norman Myers, a former British colonial officer in Africa, argued that when all species of plants, animals, and insects were included, the extinction rate would amount to at least one species per day, not one per year. Myers declared that a massive extinction of species, on a scale not unlike that of the dinosaur extinctions, was imminent (Myers 1976; Kreisler 1999). He ominously predicted that one-half of the Earth's species would disappear within the next century. The devastation he predicted would come to be called the "sixth extinction," and Myers attributed it to human activity.

Myers's outspoken publications in the 1970s and 1980s rocked the conservation community and the world of science, and he is now recognized as one of the principal architects of the contemporary conservation movement. His 1976 publication in the journal *Science,* "An Expanded Approach to the Problem of Disappearing Species," has become a foundational text of conservation science and conservation policy.

Solutions: (1) Knowledge Accumulation

Faced with a scenario of cascading extinctions, science is once again at odds with itself. One of the most prominent reactions is the claim that scientific tools—the same tools that gave us insight into the present predicament (and, some would say, contributed to it by promoting technological growth)—will prevent or halt the sequence of extinctions. The first step, according to this view, is the accumulation of scientific knowledge, submitting all species to the knowledge-building acts of observation, description, and classification.

E. O. Wilson, the Harvard social biologist (now a professor emeritus of biological science) who was among the founders of island biogeography, has become one of the most prominent voices in the contemporary call for awareness of biodiversity loss. For Wilson the need to accumulate knowledge is self-evident, requiring as its only justification the fact that species exist that are unknown to humans. Wilson stresses that our knowledge of species is in its infancy. He points out that we have only a vague idea of the numbers of species that might exist, with estimates ranging from less than 10 million to more than

100 million. Biologists have formally described fewer than 2 million of them. With the current rising rates of extinction, species are no sooner described than they are placed immediately on the Red List (Wilson 2002).

In a bold move, Wilson and six founders created the Encyclopedia of Life (EOL). The name deliberately plays with the acronym EOL, used by merchandisers to refer to a product's "end of life," the moment when it is phased out to be replaced by another. The Encyclopedia of Life constitutes a global effort launched in 2007 with a start-up pot of $50 million, including grants from the MacArthur Foundation and the Alfred P. Sloan Foundation. Its goal is to bring together the world's leading institutions—botanical gardens, libraries, and natural history museums, as well as individuals, to document all of the nearly two million species of plants, animals, and other forms of life on Earth about which we have information.

The Encyclopedia of Life organizes this information and makes it available digitally through the World Wide Web. On its website, each of the 1.8 million species has a page which grows and evolves according to the amount of knowledge documented on it. The information presented is broad—content is submitted by volunteers and authenticated by experts, and includes biological descriptions, field observations, historical literature, photos, videos, and distribution maps. The EOL is a compendium of knowledge—the enterprise of knowing life itself.

The purpose of this digital compendium is conservation. "One of the greatest assets of the EOL will be its ability to help in the conservation of our rapidly dwindling biodiversity," asserts the first issue of its newsletter.

> The current extinction rate of plant and animal species is around 1,000 times faster than it was in pre-human times—and this will increase to 10,000 times faster by the year 2050. With animals and plants disappearing more quickly than scientists can discover and study them, the EOL will serve as a vital conservation tool. Currently, there are approximately 1.8 million known and named species in the world, and scientists say there might be up to 100 million that have not been discovered. The more we know about species the harder we can work to save them. (Encyclopedia of Life 2007)

The EOL exemplifies the push to intensify zoological exploration, driven by a sense of urgency over vanishing environments (Wilson 2002). Knowledge, however, is not valued merely as an end in itself. It is an instrument to serve the ultimate goal of preserving species; these, in turn, serve the needs and pleasures of humankind. Knowledge is but the first step in facing the challenges of extinction. The second step is harvesting this diversity of species for human use.

Solutions: (2) The Pragmatic Purposes of Species: Mankind's Patrimony

Why should we save species? Much of the contemporary conservation discourse is taken up with discussions of the reasons and justifications for doing so.

In Wilson's writings we find excellent expression of the principal values of the scientific enterprise in species preservation. "Each species . . . down the roster of ten million or more still with us—is a masterpiece. The craftsman who assembled them was natural selection, acting upon mutations and recombinations of genes, through vast numbers of steps over long periods of time. Each species, when examined closely, offers an endless bounty of knowledge and aesthetic pleasure. It is a living library" (Wilson 2002:131). According to Wilson, the value of species, then, lies in two things they offer to human beings: scientific knowledge and aesthetic pleasure.

Myers, like Wilson and many other defenders of nature, justifies his concern with species loss on the grounds that it is detrimental to human well-being. He points out the many benefits of "species," a category of which humans are not a member, to human well-being and advancement: "Conservation of threatened species serves pragmatic purposes of immediate value. Genetic reservoirs make a significant contribution to modern agriculture, to medicine and pharmaceuticals, and to industrial processes in all parts of the world—especially in the advanced world, with its greater capacity to exploit genetic resources for a wide variety of purposes . . . Despite limited knowledge about genetic reservoirs, it seems a statistical certainty that tropical forests contain source materials for many pesticides, medicines, contraceptive and abortifacient agents, potential foods, beverages, and industrial products" (Myers 1976:200). In the same work he says, "There is benefit in maintaining genetic diversity not only among species but within species. I believe that we should keep as many options open as possible until, through research, we can reduce the areas of uncertainty. A principal conservation need is to set aside sufficient representative examples of biotic provinces to extend protection to entire communities of species" (198). Readers will recognize that the last statement anticipates the "hotspot" approach to conservation, one of the principal strategies of the contemporary conservation movement.

In Myers's view, humankind is not within the spectrum of nature, but rather outside it; humans manipulate and use nature to their advantage. He refers to nature as "the patrimony of mankind." As the owners of species, humans should ensure their survival in order to benefit from their utility. "If present land-use trends continue," Myers writes, "and unless better conservation measures are implemented, society stands to lose a substantial part of its heritage in species and genetic resources within a few decades" (1976:198). His position is clear: "Species can be considered an indivisible part of society's heritage now and forever" (200). His concern is that "everybody's heritage is treated as nobody's business. However much the community may regard species as its estate, it has no effective way to express this interest through institutional devices such as ownership . . . a species' intrinsic value is indivisible; if the tiger (like the atmosphere and the oceans) brings benefit to one person, it brings benefit to all" (200).

Many conservation scientists profoundly believe in the power of science to relieve us of the problems they have identified, first through exploration and description and then through application and use. In his book *The Future of Life*, E. O. Wilson argues that a myriad of resources useful to humans await our discovery. Indeed, he says, we have "only begun to explore life on Earth" (2002:14). For instance, "only a tiny fraction of biodiversity has been utilized in medicine." (119). The search for medicines and other important assets that may be found in nature is, in Wilson's words, "a race between science and extinction" (123).

> We, Homo sapiens, have arrived and marked our territory well. Winners of the Darwinian lottery, bulge-headed paragons of organic evolution, industrious bipedal apes with opposable thumbs, we are chipping away the ivorybills and other miracles around us. As habitats shrink, species decline wholesale in range and abundance. They slip down the Red List ratchet, and the vast majority depart without special notice. Being distracted and self-absorbed, as is our nature, we have not yet fully understood what we are doing. But future generations, with endless time to reflect, will understand it all, and in painful detail. As awareness grows, so will their sense of loss. There will be thousands of ivory-billed woodpeckers to think about in the centuries and millennia to come. (Wilson 2002:5)

Wilson analyzes the predicament in the metaphor of investment strategies:

> What humanity is inflicting on itself and Earth is, to use a modern metaphor, the result of a mistake in capital investment. Having appropriated the planet's natural resources, we chose to annuitize them with a short-term maturity reached by progressively increasing payouts. At the time it seemed a wise decision. To many it still does. The result is rising per-capita production and consumption, markets awash in consumer goods and grain, and a surplus of optimistic economists. But there is a problem: the key elements of natural

capital, Earth's arable land, ground water, forests, marine fisheries, and petroleum, are ultimately finite, and not subject to proportionate capital growth. Moreover, they are being decapitalized by overharvesting and environmental destruction. With population and consumption continuing to grow, the per-capita resources left to be harvested are shrinking. The long-term prospects are not promising. Awakened at last to this approaching difficulty, we have begun a frantic search for substitutes. (2002:149–150)

Wilson holds that rational bioprospecting for human improvement is the way of the future. Wilson, like Myers, favors more informed and improved bio-prospecting. They regard the transformation of nature into consumables to meet the growing needs of humans as given and unchallengeable. In these arguments, needs are biologically rather than socially determined, and therefore not within the scope of negotiation.

Solutions: (3) Space Prospecting and Settlement: A Multiplanet Species

There are abundant signs of the biosphere's limited resiliency. If the Earth is an island with shrinking biodiversity, a practical solution for a dominant species seeking to increase its prospects of survival is to transcend barriers to seek other habitats. This strategy is a prominent theme in Darwin and well documented by the island biogeographers. If humanity continues to cause growing rates of extinctions, making life on Earth impractical, as is generally acknowledged, the best option, some would say, is to migrate to seek resources elsewhere. A realm within humanity's grasp, still underexploited, is the solar system.

Carl Sagan, the astronomer and popular educator who coined the phrase "spaceflight or extinction," advocates bases and homesteads on asteroids, moons, comets, and planets to safeguard our species. In his best-selling book *Pale Blue Dot,* Sagan writes, "I urge that, with full knowledge of our limitations, we vastly increase our knowledge of the Solar System and then begin to settle other worlds . . . If our long-term survival is at stake, we have a basic responsibility to our species [to] venture to other worlds . . . When we first venture to a near-Earth asteroid, we will have entered a habitat that may engage our species forever. The first voyage of men and women to Mars is the key step in transforming us into a multiplanet species. These events are as momentous as the colonization of the land by our amphibian ancestors and the descent from the trees by our primate ancestors" (Sagan 1994:312, 332).

The physicist Stephen Hawking is a principal spokesperson for a growing number of scientists and laypersons who favor human colonization of space for purposes of survival. "If one is considering the future of the human race,"

Hawking said in 2008, "we have to go there ourselves . . . The human race has existed as a separate species for about two million years . . . If the human race is to continue for another million years, we will have to boldly go where no one has gone before." (Many may recognize in these statements the opening lines of the TV series *Star Trek,* in which the mission of the starship *Enterprise* was "to explore strange new worlds; to seek out new life and new civilizations; to boldly go where no man has gone before!")[3] Hawking says that the attempt to do so "will completely change the future of the human race and may determine whether we have any future at all" (Bardi 2008).

A great deal of scientific research and resources are already dedicated to exploring extraterrestrial alternatives to avoid extinction.[4] The unofficial race to explore, exploit, and colonize space has begun. Researchers agree that the most practical sites for colonization in the solar system are likely to be the Earth's moon, Mars, and several moons of Jupiter and Saturn. Proponents of a new space flight program hope that it will establish a moon base in the early part of the twenty-first century and land humans on Mars by mid-century.

SMART-1, launched in 2003, was the first European spacecraft to travel to and orbit around the Moon. Its robotic probe gathered data until 2006, when it was deliberately crashed into the Moon's surface to end its mission. In 2007 the Chinese National Space Administration (CNSA) and the Japanese Aerospace Exploration Agency (JAXA) each launched independent missions to explore the Moon. In 2008 the Indian Space Research Organization (ISRO) launched Chandrayaan-1, India's first lunar mission. All four missions successfully entered lunar orbits, and the agencies intend to launch lunar lander missions between 2013 and 2015. Japanese officials likened the Moon missions to the race for the South Pole (Xinhua News Agency 2007).

As I wrote this, NASA's unmanned Lunar Reconnaissance Orbiter (LRO) had just returned from its mission to relay information about the lunar environment as a first step toward creating a lunar outpost (http://lunar.gsfc.nasa.gov/mission.html). The LRO is expected to pave the way for future manned missions to the lunar surface. NASA's Space Settlement website (http://settlement.arc.nasa.gov) describes its program in space colonization this way:

> Someday the Earth will become uninhabitable . . . Those that colonize space will control vast lands, enormous amounts of electrical power, and nearly unlimited material resources. The societies that develop these resources will create wealth beyond our wildest imagination and wield power—hopefully for good rather than for ill. (NASA 2010)

Such an impulse to migrate to other habitats to ensure survival is interpreted by some scientists as compatible with the principles of population migration as deduced from evolutionary theory and island biogeography. If the world may be regarded as an island habitat with a diminishing food base, exploration in search of alternative habitats with abundant, as yet untapped resources would appear to be a wise evolutionary strategy. It also avoids any necessity of contemplating the causes of the deterioration of resources and finding solutions that would serve most humans rather than a few interplanetary colonists.

The venture well suits the ideological position that humans are superior, unfettered by the usual constraints on "species" belonging to a realm called "nature." In the metaphoric frontiers of knowledge, space is but another milestone awaiting conquest. By this reasoning, humans alone possess the power of knowledge to guarantee survival. The technological and scientific capacities of humans set them apart from all other living creatures. Through knowledge humans are invincible. Knowledge—unique to humans and justifying human exceptionalism—allows humanity to master inferior species and make them serve human needs. Using knowledge to feed growth, humans create an infinite path of betterment called "progress." The telos of this trajectory is the transformation of men into gods.

Conclusion

As happened in 1859 with the publication of *On the Origin of Species,* we find that scientific writings once again omit humankind from nature. In many recent discussions of "extinctions of species," humans are not among the species of nature, nor are they vulnerable to the forces that lead to species disappearance. Humans are not included as one species among others in the spectrum of biodiversity. Once again, humanity is relegated to the exceptional category, outside of the variety of forms of life. Species—and it goes without saying that humanity is not in this category—constitute the Estate of Man. As the number of "species" diminishes, we find that their survival is justified by prominent conservation scientists on the basis of their utility to humans. The positioning of humans as the unquestionable masters of life in all its variation appears to have changed little since 1859, even in discourses identified as scientific.

If an organizing principle in the reigning ideology of Western thought combines human superiority, exceptionalism, and invincibility, Darwin came up against it with his publication of *The Origin of Species.* The same anthropocentric worldview continues to frame much of the science associated with conserving species. As we have seen, many of the rationales for conserving species emphasize their service and utility to humanity. Moreover, technological solutions to threats of extinction, such as increased productivity or expansion to remote, unexplored regions with more abundant resources (including outer space), merely defer, rather than address, the problem of accelerated human-induced habitat loss.

If ideologies have no history (Marx 1998:42), it is because they are impervious to conditions and factors external to themselves. This would seem to be the case with the ideology of humanity's privileged status in the world. According to that overriding ideology, humans are not subject to historical influence—rather, they drive history. Thus, Darwin's evolutionism challenged the place of humankind in both the natural and supernatural worlds. This may explain why evolution has not replaced other, entrenched, theories of life on Earth and humankind's position in it, but continues in steep competition with them.

In considering, as we have, the role of science in society and the impact of society on science, we find that the two forms of knowing exist in dialogic association. In this exercise we find tensions inherent in the articulation and a lack of resolution between the two. Indeed, science, at least in the conservation field, appears to be strongly influenced by its audience. There is evidence, too, that the very production of science is less than "emancipated." Both Myers and Wilson, for example, justify preservation of "species" on the basis of their utility to humankind. In their discourse, humans are outside of the collectivity referred to as biological diversity. It is the role of humans to know other species, and thereby better harvest and consume them. An ideology of humanity's innate superiority and entitlement to other species is buttressed in these narratives. The recommended solutions—greater accumulation of knowledge and improved mechanisms of exploitation—fall short of success. They neglect the powerful role of ideology and the ways in which it is reproduced.

In considering the role of ideology in 1947, Max Horkheimer and Theodor Adorno wrote, "The advance of thought . . . has always aimed at liberating human beings from fear and installing them as masters" (2002:1). The omission of human beings from the category of natural species would seem to perform a service more accurately described as ideological or obfuscating. It refrains from that which might provoke fear and instead reiterates humankind's celebration of its superiority over things natural. To do otherwise would be to challenge ideology, to state the unspeakable. As in *Origin*, the silence prevails.

In the matter of extinctions science is at odds with itself: It is a free, unencumbered enterprise at the same time as it is determined and tethered by the society that both produces and consumes it. With Aditya Nigam we ask, "Does a science cease to be a science, if in its dynamism, it transforms its own problematic?" (1997:69). Foreclosure and reiteration better characterize the production of ideology than of science. Yet the prevalence of these features in discussions of biotic extinctions strongly blurs the distinctions between the two and lends basis for the anchoring of science in ideology.

NOTES

1. Among these are the Durrell Wildlife Conservation Trust and the Jersey Zoological Park, founded by Gerald Durrell.
2. These include the Center for Applied Biodiversity Science (CABS) at Conservation International, BirdLife International, and the Ocean Conservancy.
3. This was the phrasing in the original *Star Trek* (1966–69). In *Star Trek: The Next Generation* (1987–94) the line was changed to "where no one has gone before," which is the version Hawking quoted.
4. "Extraterrestrial alternatives" refers to two trends. For one, some authors are advocating colonies in space as refuges to thereby avoid the extinction of the human species. At the same time, efforts are underway to mine outer space for more resources as humans continue to deplete them on Earth.

REFERENCES

Althusser, Louis
    1971 Ideology and Ideological State Apparatuses. *In* Lenin and Philosophy and Other Essays, Part 2. Pp. 127–224. London: New Left Books.

Bardi, Jason Socrates
    2008 Stephen Hawking Renews Call to Colonize Space. Inside Science, April 23. American Institute of Physics.

Barney, Brett
    2006 Walt Whitman: Nineteenth-Century Popular Culture. *In* Companion to Walt Whitman. Donald D. Kummings, ed. Pp. 233–256. Malden, MA: Blackwell.

Bradford, Phillips Verner, and Harvey Blume
    1992 Ota Benga: The Pygmy in the Zoo. New York: St. Martin's Press.

Darwin, Charles
    2010[1859] On the Origin of Species by Means of Natural Selection, or, the Preservation of Favoured Races in the Struggle for Life. Madison Park [Seattle WA]: Pacific Publishing Studio.

Davis, Frederick Rowe
    2007 The Man Who Saved Sea Turtles: Archie Carr and the Origins of Conservation Biology. Oxford: Oxford University Press.

Diamond, Jared
1974 Colonization of Exploded Volcanic Islands by Birds: The Supertramp Strategy. Science 184:803–806.

1975 The Island Dilemma: Lessons of Modern Biogeographic Studies for the Design of Natural Reserves. Biological Conservation 7:129–146.

Eagleton, Terry
1991 Ideology: An Introduction. London: Verso.

Encyclopedia of Life
2007 Quarterly Quote. Encyclopedia of Life Newsletter 1:1. http://www.eol .org, accessed November 5, 2010.

Epstein, Charlotte
2006 The Making of Global Environmental Norms: Endangered Species Protection. Global Environmental Politics 6(2):32–54.

Giddens, Anthony
1979 Central Problems in Social Theory. London: Macmillan.

Horkheimer, Max, and Theodor Adorno
2002[1947] Dialectic of Enlightenment: Philosophical Fragments. Edmund Jephcott, trans. Stanford: Stanford University Press.

Kreisler, Harry
1999 The Journey of an Environmental Scientist: Conversations with Norman Myers. Conversations with History: Institute of International Studies, UC Berkeley. http://globetrotter.berkeley.edu/people/Myers, accessed November 1, 2010.

Lappé, Frances Moore
1991[1971] Diet for a Small Planet. New York: Ballantine.

MacArthur, Robert H., and Edward O. Wilson
1963 An Equilibrium Theory of Insular Zoogeography. Evolution 17:373–387.
1967 The Theory of Island Biogeography. Princeton: Princeton University Press.

Marx, Karl
1998[1888] The German Ideology. Amherst, NY: Prometheus Books.

Myers, Norman
1976 An Expanded Approach to the Problem of Disappearing Species. Science, n.s. 193(4249):198–202.

NASA
2010 Space Settlement Basics. http://settlement.arc.nasa.gov/Basics/wwwwh.html, accessed March 19, 2011.

Nigam, Aditya
1997 Marx, Althusser and the Question of "Power," in Lieu of a Rejoinder. Social Scientist 25(3–4):65–71.

Olalquiaga, Celeste
2006 Object Lesson/Transitional Object. Theme issue, "Ruins," Cabinet Magazine 20. http://www.cabinetmagazine.org/issues/20/olalquiaga.php, accessed November 5, 2010.

Phillips, Adrian
2004 The History of the International System of Protected Area Management Categories. Parks 14(3)4–16.

Quammen, David
1997 Song of the Dodo: Island Biogeography in an Age of Extinction. New York: Touchstone.

Sagan, Carl
1994 Pale Blue Dot: A Vision of the Human Future in Space. New York: Random House.

Simberloff, Daniel, and Edward O. Wilson
1969 Experimental Zoogeography of Islands: The Colonization of Empty Lands. Ecology 50(2):278–296.
1970 Experimental Zoogeography of Islands: A Two-Year Record of Colonization. Ecology 51(5):934–937.

Wilson, Edward O.
2002. The Future of Life. New York: Knopf.

Wilson, Edward O., and Daniel Simberloff
1969 Experimental Zoogeography of Islands: Defaunation and Monitoring Techniques. Ecology 50(2):267–278.

Xinhua News Agency
2007 Mission to Moon Not a Race with Others. Space Daily, August 20.

Yanni, Carla
1999 Nature's Museums: Victorian Science and the Architecture of Display. Baltimore: Johns Hopkins University Press.

2. FROM ECOCIDE TO GENETIC RESCUE

CAN TECHNOSCIENCE SAVE THE WILD?

Tracey Heatherington

Banking on Extinctions

A few years ago, the late Stephen M. Meyer announced "the End of the Wild" in the *Boston Review*. A distinguished MIT professor and passionate advocate for environmental policy, Meyer told us we had already lost all chance of saving "the composition, structure and organization of biodiversity" in nature (2004:1). His grim assessment was supported by the Fourth Assessment Report of the Intergovernmental Panel on Climate Change (2007), which pronounced it "virtually certain" that the rates of species loss will accelerate due to trends in anthropogenic global climate change. Recently, the Secretariat of the Convention on Biological Diversity (SCBD) rather bluntly acknowledged that an eight-year international program intended to stem the tide of extinctions by 2010 failed to achieve its goal.[1] Projections suggest significant impending changes in the distribution and abundance of species and biomes over the twenty-first century (Leadley et al. 2010).

These unprecedented levels of species extinctions result not only in physical losses, but also in losses to traditional knowledge systems, future scientific knowledge, and the human spirit. Possibilities of better understanding human history itself are being lost along with the wild. In the years since the first Rio Earth Summit, the growing role of genomic research in environmental conservation science has promised new visions of nature and the way we should manage it. In 2004, for example, the Natural History Museum of London, in collaboration with the Zoological Society of London and the Institute of Genetics at the University of Nottingham, launched an initiative called the Frozen Ark (Natural History Museum 2004).[2] The project collects DNA samples and viable cells from endangered species and freezes them at minus 80°C. One journalist proclaimed it "a modern version of Noah's Ark, designed to save thousands of creatures from extinction." He explained, "Scientists . . . are keen to preserve the DNA of endangered animals so they can continue research into their evolutionary histories even if they become extinct. More ambitiously, scientists hope one day to be able to use cells from the frozen tissue samples to recreate extinct animals using advanced cloning techniques" (Sample 2004). A growing international consortium of institutes, zoos, and museums around the world have signed agreements to store and share materials and information in this "DNA bank."[3] The institutional partnerships, financial articulations, technoscientific capacities, and global imaginaries mobilized for the Frozen Ark define an important new "global assemblage" (Ong and Collier 2005; Collier 2006). They suggest that gene banking and cloning biotechnologies will play a crucial role in the future of biodiversity conservation.

With its biblical reference to imminent world disaster and its transcendent faith in technoscientific interventions, the Frozen Ark project reflects the moral discourses of global environmental movements that, after the multiple failures of the initiatives envisioned at Rio, are now permeated with urgency and irony. The Frozen Ark rescues representative samples of genetic diversity in icy vessels of liquid nitrogen, so that we may one day renew the abundance of the Earth in a future graced by better knowledge and moral understanding of both ecological systems and the foundations of life itself. The Frozen Ark is made feasible by burgeoning investments in genomics and emerging techniques associated with somatic cell nuclear transfer, usually glossed as "cloning." Since this is a branch of science that has generated vehement reactions from fundamentalist religious groups, the biblical allusion of the Frozen Ark is both resonant and strategic across much of the Western world. Here at the frontier of imaginary futures, the moral and symbolic worlds of the Old Testament, environmentalism, and genomic science awkwardly converge. Will our growing ability to intervene in nonhuman systems of reproduction now offer redemption for the role humans have played in species extinctions? If so, how will this alter the nature and culture of biodiversity protection?

The implications of the last question evoke Paul Rabinow's (1996) famous prediction that we have entered an era of "biosociality" in which the social imagination will reshape the foundations of planetary life and biological experience, naturalizing its own interventions in the process. With the advance of new reproductive technologies and a growing ability to both understand and manipulate human and animal genomes alike, unexpected evolutionary vistas beckon. Against the bleak tide of species losses, the gleaming possibility of new science and technology gathers force. This essay considers "the end of the wild" in relation to the moral and epistemological reframings of biosociality in the twenty-first century. Using an example from Sardinia, I offer a cultural perspective on initiatives for genetic banking and assisted reproductive technologies (ARTs) applied to nonhuman species.

As an anthropologist, I am concerned that the objectification of endangered species as genetic resources tends to decontextualize them, removing them from locally embedded and culturally meaningful patterns of human-animal relations. This is a process already far advanced by mechanisms of globalization, which jeopardize traditional subsistence economies and cultures, and by wilderness conservation movements, which often privilege romantic Western visions of untouched nature or naïve ecological stewardship over nuanced understandings of traditional cultural ecologies. Projects to "bank" and "retrieve" the DNA of endangered species have generated a public discourse with strong overtones of what anthropologists call "genetic essentialism." Intense public interest in the human genome project, for example, has tended to reinforce conventional ideas that social identity is defined by an innate biological "nature" perceived to be coded in DNA (Simpson 2000; Heller and Escobar 2003).[4] As genetic and genomic sciences breed new efforts in the fields of both biological prospecting and biodiversity conservation, the very mission of wildlife conservation is apt to be equated with the conservation of genetic resources. This steals respectful attention away from local ways of seeing, perceiving, and embodying relationships with landscapes and nonhuman species. The cloning of a mouflon sheep indigenous to Sardinia, a Mediterranean island in Italy, highlights the genetic essentialism implicitly tied to public discussions of DNA rescue and the assisted reproduction of such wild species. These strategies give moral authority to scientists and conservation experts, while obscuring the importance of Sardinian cultural tradition. "Sheepwatching" (Franklin 2001) in Sardinia inspires fertile narratives about the authenticity of nations and wildlife, cross-species reproductive technologies, and the contested morality of animal salvation.[5]

The Moral ARTs of Biodiversity Retrieval

Considerable excitement surrounded the first successful animal cloning experiment in 1996, when Dolly the sheep was brought into the world at the Roslin Institute near Edinburgh. Dolly was artificially conceived from genetic materials contained in an adult somatic cell. Her existence implied revolutionary possibilities for the production of food, pharmaceuticals, and webs of significance associated with animal breeding.[6] As Sarah Franklin (2003) discusses, cloning Dolly represented a transformation of reproductive potential that implied fundamental changes in the way we understand our bonds of both kinship to and ownership of domesticated species. Franklin explores how intellectual property law, market forces, gender, and science now converge to naturalize genetically engineered brands as emergent forms of paternity. We might say that cloning the domestic sheep was a "seminal change" in the science and practice of animal husbandry. It also set the scene for an interconnected and symbolically charged story about Ombretta the mouflon, another cloned sheep. What Franklin (2003, 2007) calls Dolly's biological and commercial "viability" has heralded implications for conservation biotechnology in the age of genomics. Several projects to clone endangered animals took off between 1998 and 2000. Within the larger ambit of conservation science referred to as "genetic rescue," assisted reproduction is now viewed as one of a set of tools that can be used to maintain genetic diversity within threatened species populations. Cloning techniques could represent an important option where animals threatened in their indigenous habitats do not breed well in captivity. Endangered species cloning projects are controversial and high-profile. Focusing on popular megafauna, they are symbolically important for the nation-states, public entities, and private institutes that invest in them. "Sheep-watching" in Sardinia suggests fertile narratives about the authenticity of nations and wildlife, cross-species reproductive technologies, and the contested morality of animal salvation.

In the spring of 2001, a collaboration between the Roslin Institute and a research team from the University of Teramo, Italy, made the first surviving clone of an endangered mammal: a subspecies of the mouflon, *ovis orientalis musimon,* a wild sheep indigenous to the Mediterranean islands of Sardinia and Corsica (Istituto Regionale Foreste Sardegna 2006a, 2006b). A *Nature Biotechnology* article announced that, by means of the same technique of somatic cell nuclear transfer once used to create Dolly, genetic material had been "rescued" from two dead Sardinian mouflon ewes and used to replace nuclei in the eggs of a closely related domestic sheep species (Loi, Ptak, et al. 2001). Seven viable embryos were obtained and implanted in surrogate domestic ewes. One lamb was carried to term, was birthed, and survived. Journalists reported that the lamb, named Ombretta ("little shadow" in Italian) after the dead mouflon from which scientists harvested the DNA, was "repatriated" to a wildlife rescue center on the island of Sardinia. She was important not only as living proof that cloning science could be applied to biodiversity conservation strategies, but also because her DNA constituted a viable "resurrection" of genetic material from a deceased animal. Because her genetic material was received from a mouflon donor, she was considered an authentic reproduction of a "wild" animal. The scientists who cloned Ombretta suggested that "endangered wild animals could perhaps be salvaged or maintained using the oocytes and wombs of their unthreatened domesticated counterparts . . . Cloning has the potential to preserve, and even expand, genetic variability" (Loi, Ptak, et al. 2001:962).

Amidst fierce debates on the ethics of cloning, this application of ARTs to wildlife conservation appeared relatively felicitous. While journalists were quick to note Dolly the sheep's commercial value to big agriculture, no one immediately branded the baby mouflon with the ignoble title of "cash cow." The marketable applications for transgenic cloning techniques in domesticated species include the reproduction of valuable purebred lines of cattle, horses, and dogs, and "pharming" medicines and even organs for human transplant from genetically engineered animals. These are not at issue for the successors of Ombretta. Although intellectual ownership of innovative cloning techniques is commercially valuable, the returns on cloning experiments for endangered species are generally lower than for domestic species, given laws that make it illegal to breed them for sale or slaughter. Since the DNA samples available for long-extinct species are generally too degraded to support cloning science, public fears of *Jurassic Park*–style enterprises have faded with the popularity of these films. Instead, objections to the cloning of endangered animals have focused on animal welfare issues associated with the high incidence of aborted pregnancies and the low viability and accelerated aging of cloned offspring. Critics have also worried that popularizing the idea of cloning as a technological fix to biodiversity loss could derail efforts to establish nature reserves, protected areas, and wildlife corridors, which are needed to conserve habitats and ecosystems more holistically. They have pointed out that cloning individuals may do little to support genetic diversity within species, and experimental forms of genetic engineering could put existing species populations at risk if the cloned individuals pass on genetic flaws to their natural offspring.[7]

Given the scarcity of resources for environmental conservation, cloning is generally considered too expensive and inefficient to be the best means of rescuing species at risk today. Scientists working on these projects emphasize that cloning techniques must be part of a larger strategy to rebuild and protect genetically viable populations of species under threat. Enormous investments have been made in a handful of high-profile biodiversity cryogenics and cloning projects. Prominent scientists working at Advanced Cell Technologies have written, "Although we agree that every effort should be made to preserve wild spaces for the incredible diversity of life that inhabits this planet, in some cases either the battle has already been lost or its outcome looks dire." They believe that "some countries are too poor or too unstable to support sustainable conservation efforts" and that cloning technologies offer an economical alternative to maintaining animals in captivity (Lanza et al. 2000:85, 89). These authors advocate Frozen Zoo–style initiatives. "The truth of the matter is," said Robert Lanza in an interview, "if an animal is killed right now, the genetics of that animal is lost from the planet forever . . . [but now technology exists so that] when an animal dies, all you have to do is freeze a few cells to preserve the genetics of the animal forever" (Lanza 2002:8–9). Dr. Betsy Dresser, director of the Audubon Institute Center for Research on Endangered Species (AICRES), insists that "the general public tends to be unaware of the good that this technology can do. If I have to choose between extinction and cloning, I'll choose cloning every day. Extinction scares the devil out of me" (Dresser 2001).

The International Union for Conservation of Nature (IUCN) is one of the partners supporting the Frozen Ark initiative, and the Audubon Nature Institute runs its own cryogenics and cloning lab near New Orleans. The Frozen Ark collects genetic material from threatened species of all kinds. However, projects to clone endangered and extinct species have focused exclusively on "charismatic megafauna,"[8] including the South Asian gaur, the Sardinian mouflon, the Banteng cow, the South African wildcat, the Chinese panda, the Asiatic cheetah, and the extinct Tasmanian tiger.[9] As key symbols of national histories and identities, with aesthetic and emotional appeal cultivated by preexisting discourses of wildlife protection, these mammals command the focus of the social imagination and incite sympathy for nature conservation. This has certainly been true of the Sardinian mouflon, which the Sardinian Forestry Service calls "the animal most representative of the island" (Istituto Regionale Foreste Sardegna 2006a). Yet when Ombretta the cloned mouflon was brought back to the island where her genetic predecessor had once lived, her Sardinian citizenship was constructed through metaphors that conflated biological heritage with indigeneity.

Cloning the Wild Mouflon

The mouflon lives both singly and in small herds in the rocky, mountainous countryside of Sardinia. It jumps and climbs with agility, and can run about sixty kilometers per hour. Although populations of mouflon were brought to the Alps and Apennines, where they are not considered under threat, the population native to Sardinia dwindled throughout the twentieth century due to hunting, loss of habitat, and competition from domestic sheep. They are protected by the Bern convention (1979), the EEC Directive on the Conservation of Natural Habitats and of Wild Flora and Fauna (1992) and the Sardinian wildlife protection law (1998). The herds have been monitored and managed by the Sardinian Forestry Service since the 1980s, and by 2006 their numbers had revived to approximately 3000 (Istituto Regionale Foreste Sardegna 2006b). The largest concentrations of mouflon are found in the region of the Gennargentu mountains and the Supramonte (east-central Sardinia), where the harsh landscapes have not been developed for intensive agriculture, tourism, or industry. Most such areas either have remained under the control of the Sardinian Forestry Service or are owned corporately by residents of traditional herding towns and used for extensive natural pasture, hunting, and gathering. Fierce local attachment to common property systems as the basis for pastoral production has supported anthropogenic landscapes that are often mistaken for wilderness. These landscapes have been home to the mouflon.

Pastoral activities continue to flourish alongside forestry and ecotourism in the Sardinian interior, where many herders still milk their animals without machinery and produce hand-crafted cheeses. Pastoralism is vital to local identity and feeling, both as a historical experience and as a current strategy to support economic security and well-being in the face of difficult and changing circumstances. Pastoral traditions are a central focus of ethnonationalist pride in Sardinia as a whole. Yet conservation experts often consider pastoral activities to be a key source of environmental degradation and conflict with wild species over available resources.[10] It must be said that the rural, highland region of Sardinia is widely perceived to be a place—as well as a people and culture—set apart, different, and perhaps even fundamentally troublesome. In Sardinia we find investments made to promote both nature and culture tourism in a region where development policy has long sought to transform and modernize systems of transhumant herding, in conjunction with antikidnapping and antibanditry campaigns.

As a rural frontier reminiscent of the American "Old West," central Sardinia occupies a poignant place in the public imagination. Encapsulated within a "modern" Italy and a dynamically progressive European Union, central Sardinia is perceived to bear the marks of a "less developed" economy. National inquests into Sardinian banditry in the late 1960s retraced nineteenth-century discourses linking race, cultural essentialism, and "backwardness."[11] These have carried forward in debates about anthropogenic forest fires, poaching, theft, kidnappings, political intimidation, and high rates of homicide that continue to plague the poorest communities of highland central Sardinia. From the 1960s until 2005, the World Wide Fund for Nature (WWF) campaigned to "save the mouflon" by establishing a national park in this area, hoping to consolidate municipal areas of communal pasture under a centralized management authority that would restrict both traditional land use and development schemes. The first legislation intended to establish the controversial Gennargentu National Park was put aside after memorable rural demonstrations against it in 1969. In 2005, protesters from Sardinian pastoral towns marched upon Cagliari, the regional capital, demanding that legislation approving the national park in 1998 be rescinded. Environmental advocates have claimed that the political opposition of rural Sardinians to the national park confirms their stubborn ignorance, traditional orientation, and lack of environmental values. Themes of cultural essentialism and backwardness reappear frequently in interpretations by the media, prominent NGOs, and some urban bureaucrats of local resistance to the creation of a national park (Heatherington 2010).

Pictures of the cloned baby mouflon put Sardinia on the map for nature lovers and technology enthusiasts around the world. Yet these pictures are not innocent; they evoke a complex history of contested political ecology and a future imaginary of technoscience and environmental redemption. On an island of traditional herders, endangered wild species and their habitats are increasingly valued over the semidomestic animals that constitute the fundamental basis of pastoral wealth and heritage. The irony that traditional agro-pastoral systems protected the very biodiversity that multiple levels of legislation now intervene to preserve is not lost upon rural Sardinians. Hence, the mouflon has been reappropriated as an iconic image of irony by Sardinians seeking to intervene in the politics of conservation. Placards written for the 1969 campaign of opposition to the Gennargentu National Park cried out against policies that privileged goals of wildlife conservation over local development. "You save the mouflons before the men," they announced. More than 35 years later, on a banner carried by townspeople from Urzulei in the demonstration at Cagliari in 2005, the president of Sardinia was portrayed as a "protected animal" with the body of a mouflon. In 2006, the same president announced that he had no further plans to implement the legislation to create the Gennargentu National Park, given the ongoing lack of local support. Other species conservation initiatives continue.

Communities that once fostered sustainable agro-pastoral systems are increasingly excluded from biodiversity management. Today, the world's largest environmental nongovernmental organizations actively participate in the management of Mediterranean biodiversity through a series of partnership agreements with government agencies and international organizations. They see Sardinia as part of a distinctive Mediterranean ecological zone with significant biodiversity resources under threat (see, for example, Cuttelod et al. 2008). WWF is particularly active in efforts to protect Italian wildlife and habitats. Discourses of development, criminality, and environmental management continue to present Sardinian pastoral traditions as a block to progress. Central Sardinia has been stereotyped as a resistant frontier, to which the cosmopolitan science of biodiversity "hotspots" and "ecoregions"[12] must be brought to enlighten narrow-minded locals and convert them to the moral ecology of ecodevelopment. Any real possibility that local knowledge and practice could help guide efforts at conservation and sustainable development is suppressed as a result of this attitude.

Policy debates and public controversy over a new national park in Sardinia present a vivid backdrop for the scientific innovations that won global attention in the cloning of a wild mouflon. Environmental politics here are suffused with traces of essentialism. Visions of a backward, somewhat dangerous, autochthonous culture have mingled and merged with romantic visions of rugged wilderness, hidden in the heart of Mediterranean Europe. It is this seeming convergence of "wild culture" and "wild nature" that makes Sardinian genetic legacies an object of profound interest to both science entrepreneurs and the nation-state. The very idea of "repatriating" genetic resources embodied in a laboratory lamb reconfirms the rootedness of assumptions about genetic essentialism in the public imagination and in policy discourses.[13] After all, the inherent "naturalness" and "wildness" of Sardinia legitimates narratives of sovereignty and distinction, attracts attention to biodiversity, and creates new markets for the consumption of authenticity in an age of endangered autochthons. Given the early confluence of anthropometric studies and criminology in Sardinia, the troublesome authenticity of pastoral culture is still often taken to be associated with biological roots, so that a misleadingly simplistic version of genetic essentialism tends to persist subtly in the public domain.

It is provocative to note that about the same time that the mouflon cloning project was underway, genomic research proceeded on a small group of people in rural highland Sardinia. Biological anthropological studies in the Ogliastra region (where several localities agreed to participate in Sardinia's Parco Genos Consortium) used mitochondrial DNA analysis in an attempt to map patterns of genetic homogeneity for complex traits.[14] The Ogliastra was treated as a geographically restricted and isolated area with a high frequency of genetic endemism. This is a striking parallel to both the nineteenth-century anthropometric studies in highland Sardinia and the ongoing efforts of conservationists to preserve endemic species in the same region. Talana and Urzulei, two of the three towns chosen as the focus of human genetic research, have also been affected by the plans for the Gennargentu National Park. Thus human and animal domains of genetic research also cross-fertilize to reembed genetic essentialisms in cultural discourses—with tremendous implications for local communities whose territories are now subject to centralized biodiversity management, while the bodies of their members have become the subjects of biomedical research.

Biogenetic Futures

In the domesticated sphere of agricultural systems, the existing biodiversity of farm animals is already deeply shaped by practices of genetic testing, modification, cryopreservation, and artificial reproduction that now augment older breeding strategies. Cloning and genetic certification of domestic animal breeds are fast transforming the economic and regulatory context in which farmers work. They also deeply affect the global imagination. Consider this recent advertisement from the American company ViaGen, which provides private services to breeders:

> By gene banking your most valuable animals, you protect your investment, and preserve the option to multiply that investment later through cloning. If you lose an animal to injury or disease, or you geld a colt that later proves to have valuable reproductive potential, gene banking is sound insurance. When you order our CryoSure® service, our scientists culture your animal's cells before cryopreserving them, to confirm that the cells are viable and that you will have the option to clone the animal whenever you're ready. (ViaGen n.d.)

The future of agricultural production looked very different on ViaGen's technologically savvy 2008 website than it did recently in rural Sardinia, where, on more than one occasion, I have watched men milk their animals by hand. This difference in perspective is important.

What does Dolly's "wild" cousin, the cloned mouflon lamb, signify for Sardinia's pastoral heritage, and for the strategies of cultural authenticity that are bound to it? Whereas traditional social ecologies in Sardinia entail fluid relationships that weave together whole, dynamic ecosystems in cultural perspective, the dominance of technoscience in institutional assemblages for biodiversity management tends to isolate static and highly valued objects of conservation in material and symbolic economies. The global celebration of Ombretta's birth highlights the power of new science to reconfigure and reauthorize political authority over biodiversity, now a key resource. The interspecies cloning of endangered sheep is still far from routine, but "genetic rescue" has assumed an ongoing role in European mouflon conservation programs. The techniques involved in genetic rescue have usually centered on basic species population management, first evaluating reproductive fitness, using a combination of monitoring mechanisms and statistical approaches to measure genetic variation, and then intervening to ensure viable colonies or subpopulations.[15] Interventions most often take the form of conventional breeding programs and the reintroduction of breeding partners in wild species populations where the number of individuals has fallen critically low. In the case of the European mouflon, however, genetic management programs include a DNA resource bank with cryopreserved sperm, embryos, and somatic cells, as well as methods of in vitro embryo production and the induction of cross-species pregnancies by implanting embryos in common domestic ewes (Ptak et al. 2002).

The way that such technological strategies are imbricated in larger questions of regional or national development and governance with respect to a notorious minority in central Sardinia raises new questions about the relative values of nature and culture in a postgenomic world. If we look at endangered sheep from a population genetics perspective, for example, our attention is largely shifted away from a Sardinian cultural heritage that has historically supported a habitat for mouflon survival, toward high-level scientific cooperation and accomplishment. Even the modest path to environmental sustainability envisioned at Rio mandated support for social justice issues and the life projects of autochthonous peoples. The story of the cloned mouflon suggests that the guiding vision of biodiversity conservation is turning elsewhere. Thanks to the "future imaginaries" (Fujimura 2003) promoted by science entrepreneurs like those involved in the Frozen Ark, the public embraces the miracles of assisted reproduction and cryobiology. This embrace gathers moral agency into the hands of complex institutional assemblages that sustain scientific expertise, while it obscures what human/animal relations might look like from any other culturally rooted perspective.

The dramatic language of genetic rescue and the salvation of wild species is perhaps more relevant to public environmental advocacy strategies than to the assumptions of basic ecological science. Mundane monitoring of population health and protection of habitats is necessarily the mainstay of wildlife management for most biodiversity conservation programs. Yet the moral terrain of extinction is tremendously evocative for the genetic imagination, defining the frontiers of capital investment in both technoscience and biodiversity. With respect to other ethically laden debates on technologies for genetic research and tissue engineering, the artificial reproduction of endangered mammals through transgenic cloning techniques implies potent new forms of kinship with both wild and domestic species. The stakes here for nation-states are vividly apparent. Frontiers of science, culture, and biodiversity constitute powerful domains of identity construction as well as economic growth. Endangered megafauna, as the objects of cloning programs, legitimate and naturalize multiple narratives about sovereignty, governance, risk, and opportunity. Thus, while many little-known species of indigenous reptiles, amphibians, insects, birds, and

plants may be lost in Sardinia to climate change, development, and other anthropogenic intrusions, extraordinary investments of effort, capital, and imagination have been made to protect and bolster the numbers of mouflon.

The mouflon is now simultaneously an emblem of distinctive national heritage and identity, proof of living national wealth in biodiversity (a commodity
likely to escalate in value for ecotourism as extinction events proceed), and the
embodiment of national achievement in science and technology, with all their
implicit economic potential. As an excited rhetoric of innovation takes root
in a competitive investment environment, the symbolic resonance of the wild
mouflon redoubles in relation to its poor semidomestic cousin, the "ordinary"
sheep kept by the troublesome, not-so-modern heirs of Sardinian pastoralism.
Sarah Franklin (2001) notes that sheep and humans have a special connection within biblical narratives of sin, shame, and redemption. This connection
persists in this era of technoscience and environmental crisis. Yet in the "brave
new world" (Huxley 1989) of Ombretta, the cloned mouflon, the embeddedness of sheep in Sardinian society—that is, the role of Sardinian pastoral traditions in mouflon survival—is largely forgotten in projects of wildlife conservation. In a world shaped by both extinction events and postgenomic science, a
new generation of rural Sardinian shepherds find their marginality renewed.

Return to Eden: Immortal Values and Cultural Portfolios

In global terms, the evolution of genomic science has generated both
new risks and new possibilities associated with environmental security. One
prominent concern related to biodiversity protection is food security. Given
the growth of monocultures associated with industrial agribusiness and the
ongoing introduction of genetically modified organisms in commercially oriented farm systems, DNA banking is viewed as important to stemming the
loss of plant biodiversity in staple crops. Cary Fowler, head of the Global Crop
Diversity Trust, sees new technologies of conservation as insurance of human
survival in the face of emerging risks. In addition to supporting local seed
banks to preserve agricultural heritage in many developing countries, he has
worked to establish a permanent, centralized bank of seeds from around the
world (Fowler 2008).

In February 2008, the Global Seed Vault—popularly dubbed the "Doomsday Vault"—was opened in Svalbard, Norway. It was inaugurated by Kenyan
Greenbelt Movement leader Wangari Maathai, together with the Norwegian
prime minister, who delivered the first seed packets through a four-hundred-
foot tunnel into the tight-security storage facility in the heart of a mountain.
The seed vault is intended to enable the regeneration of agriculture in the event

Figure 2.1. Global Seed Vault, "Snow and Wind." Mari Terre/Global Crop Diversity Trust, http://www.croptrust.org/main/arctic.php?itemid=217, accessed March 22, 2009.

of a "worst-case scenario" such as nuclear war or massive crop failures resulting from natural disasters, disease, infestation, or civil insecurity. Altered conditions associated with climate change, for example, or unintended contamination by genetically modified crops could deeply undermine the global heritage of genetic diversity in seed crops. The vault was constructed approximately six hundred miles from the North Pole of the Earth, and its deep cold and remote location, together with Norway's established political neutrality and international peace agreements, are thought to provide security for the long-term cryogenic storage of heritage seed crops. "In the past, when accidents or natural disasters or war intervened and destroyed [seed] samples, then that was it—they were as dead as a dinosaur, extinct," Fowler said when his initiative was realized. "But we are going to put an end to extinction with this vault because we are going to have a safety backup, a Plan B" (Acher 2008).

The vault does not accept genetically modified seeds. This illustrates the profound human ambivalence attached to our growing capacity to transform life itself. The very technologies that enable industrial agricultural production are now acknowledged to put global food security at risk through the effects of pollution, anthropogenic climate change, and loss of crop genetic diversity. This irony is captured in the design of the beacon intended to signal the location of the seed vault. Commissioned by Public Art Norway (KORO) to design an installation of light and mirrors for the Svalbard Global Seed Vault, artist Dyveke Sanne hoped that her work, entitled "Perpetual Repercussion," would inspire reflection:

> The building will reflect the light depending on the time of year and the time of day. During winter darkness other sources of light will replace the sunlight. The light will complement the darkness that has been dug out of the permafrost soil, and will signal the seed vault's positioning at all times. The depths of the seed vault are out of sight. Yet, its contents reflect a meaning and a complexity that affects us. From the moment we become aware of its existence, we are reminded of our own position in a global perspective and the condition of our planet. The seeds carry an obligation to the future. They are copies for a diverse landscape that demands cyclical repetition of actions, rather than continued faith in the designated original and linear progression. The mirrored surfaces do not betray any underlying contents. They copy and reflect what they receive. Close up you can see yourself in the mirrors, further away you become part of the landscape, or are blinded by the reflected light. The reflections come together and shift internally according to the position of the viewer. (Sanne 2008)

The Doomsday Vault would at first seem to imply some technocentric naïveté that a return to an original state of nature may one day be possible despite impending environmental impacts of climate change or the potential devastation of nuclear war. When the vault opened, European Commission president Jose Manuel Barroso declared to attending journalists, "This is a frozen Garden of Eden!" (Mellgren 2008). This powerful metaphor of human ecological innocence evokes a theme of technomillenarian redemption for the global ruin of biodiversity. The self-conscious intention of international collaboration at Svalbard, however, is to preserve not merely seeds, but the possibility to better know and understand "the book of life" (Kay 2000) contained in DNA. Indeed, one of the great ironies entailed in the monumental vault is that the authentic genetic heritage conserved here may ultimately be used as a basis for further genetic research and modification of crops.[16] Despite the romantic imagery associated with the Global Seed Vault, what we are really trying to protect is not

Eden itself, but the proverbial apples of knowledge. We fill our pockets with them as we leave the dying garden, hoping to savor and consume them later.

Some critics of the Svalbard Seed Vault insist that local, on-farm conservation should be prioritized. Henk Hobbelink, director of GRAIN, a nongovernmental organization based in Spain that promotes agricultural biodiversity, stated that "the problem that we have with this much-publicized project is that it tends to divert attention, energy and money away from what we consider as much more urgent and sustainable efforts to save biodiversity on the farm" (Roug 2007). As Stephen Brush's (2000, 2004) extensive work on comparative farming systems affirms, ex situ seed banks may be useful, but they should not distract us from supporting the practical and ritual knowledge bases associated with indigenous agriculture in situ.[17] Like the Yunnan agriculturalists struggling to achieve recognition and support for their place-based traditions of biodiversity management (Hathaway, this volume), rural people who participate in small-scale farming and pastoral systems have much to contribute to sustainable futures. Inside the mountain at Svalbard, massive databases connect the seed packages to catalogues of information about the nature, characteristics, and uses of the plants being saved. What may not be conserved here is the deeper cultural knowledge about the ecosystems, languages, religions, and social relations that maintained this heritage of genetic diversity for millennia before the seeds arrived in the vault. The databases of agricultural biodiversity at Svalbard decontextualize genetic resources from the historical social contexts in which they were embedded, relying upon information systems that strip away sensual ephemera and the idioms of community, and inflect codified knowledge with the values and perspectives of Western science. A small handful of new initiatives to produce complementary databases of cultural knowledge about agricultural practice to support new seed-saving technologies (Nazarea 2006) represent an important missing piece of the puzzle. Yet the genetic essentialism inherent in popular ideas and enthusiasm about the global initiative still threatens to undermine the role and value of local knowledge. The Svalbard Seed Vault is designed like a military bunker and guarded like a bank vault. As an icon of environmental security, it normalizes the fortification of seed commodities, even as it fails to protect agricultural communities from crisis events or the banal indirect violence of a harshly neoliberal global economy.[18] If we devote already scarce resources to protecting genetic heritage without adequately defending the social and cultural capacity to maintain sustainable agricultural systems, our sacrifices may be wildly misspent.

Increasingly, human efforts privilege the conservation of wildlife through habitat protection and the maintenance of viable population numbers of species in the wild, as well as through technologically assisted reproduction, the cryogenic storage of DNA, and the cloning of endangered or even extinct non-human species. Professor Mike Archer, dean of science at Sydney's University of New South Wales, now famous for his work on ancient DNA, proclaimed recently, "There used to be a time when extinction meant for ever, but no more" (Shears 2008). Without doubt, the future biological evolution of "the wild" will take paths ever reshaped by human social perception, intervention, and indifference. At the same time, advancing biotechnology and its applications have provoked a gestalt shift in conceptual and practical approaches to agriculture. We imagine immortal values attached to genetic codes that can be frozen in time and transcribed in advanced information systems. An uncanny combination of biblical and financial investment metaphors animate these virtually disembodied life forms. Today, biosociality affects not only the human condition but all of life on Earth. In the twenty-first century, biogenetic futures for human and nonhuman species alike are distinguished, patented, commodified, traded, deposited, saved, coded, networked, and continually modified (see Hayden 2003; Bowker 2005; Rajan 2006; Franklin 2007; Pálsson 2007).

The anthropogenic destruction of natural and traditional ecological systems—sometimes referred to as "ecocide"—has been linked with both ethnocide and genocide. For example, environmental destruction, contamination, and injustice jeopardize the cultural survival of autochthonous groups still living from the land across Amazonia, Indonesia, the Arctic, the Kalahari, Western Australia, Papua New Guinea, and the small island states of the Pacific. Remaining undeveloped spaces and resources are rapidly being inventoried, commodified, privatized, and commercialized under the guise of development, bioprospecting, and even ecotourism. Local cultural practices are marginalized, displaced, or transformed into performances for outsiders. A wealth of indigenous ways of knowing, recalling, and being-in-the-world are forgotten as cultural diversity is eroded. As Bernard Perley points out (this volume), these "collateral extinctions" represent the loss of living landscapes, in which the rooted connections between particular communities and natural biodiversity are inherent in language and other forms of intangible cultural heritage, as well as social experience. Until we truly invest in supporting and revitalizing these living connections to place and nature, biodiversity may be impoverished by our very notions of how to "bank" and "retrieve" it.

Arguably, our quest to rescue and preserve nonhuman biodiversity under the auspices of technoscience entails a particular cultural way of seeing that denies its own ethnocentricity. To consider knowledge about nature from the "modern" standpoint is to presume that legitimate knowledge can be derived only from observation of the tangible, objective world, the world that can be known only through science. The knowledge produced by scientific experts overshadows all other forms of intangible cultural knowledge about the past. We "naturalize" a monoculture of knowledge, vision, and understanding. A paradox emerges: even as we move to stem the loss of biodiversity that endangers our ability to understand indigenous and natural histories, we find the authenticity of wild nature, and everything we may learn from it, compromised by our own technocentric epistemology. What we preserve only reproduces our inclination to biological essentialism and functionalist values. It does not protect the "wild profusion" of dynamic, mutually embedded, natural and cultural forms of life we find in situ (Lowe 2006). If we think of extinctions only in terms of lost nature, we fail to identify and stand against forces that diminish the world's cultural portfolio of living landscapes. Even though we save genetic diversity as the embodiment of potential knowledge about the past and something instrumental to the future of humanity, we may nevertheless forget that it is possible to envision history, nature, and the time to come in multiple ways. "The end of the wild" will signify epistemic erasures of desperate magnitude, as culturally embedded pasts and futures slip beyond memory and imagination.

ACKNOWLEDGMENTS

This paper is a revision, update, and expansion of "Cloning the Wild Mouflon," an article I published in 2008 in *Anthropology Today* 24(1):9–14. The first four sections of this text are reprinted with permission from Wiley-Blackwell. Thanks to Bernard Perley, Genese Sodikoff, Peter Whiteley, and David Hughes for comments on recent versions. I am also indebted to Pasqualino Loi for his gracious encouragement.

NOTES

1. The Convention on Biological Diversity (CBD), signed at the 1992 UN Earth Summit at Rio de Janeiro, inaugurated international efforts to collaborate and govern biodiversity as a global commons. The SCBD oversees the implementation and assessment of ongoing agreements.

2. In 2000, a parallel project called the Millennium Seed Bank was established by the Royal Botanical Society in Britain to safeguard genetic diversity in plants. The project assembles information about seed varieties and preserves seed samples for long-term storage through a process of drying and freezing. Like the Frozen Ark, the Millennium Seed Bank project assembled partnerships around the world. This seed bank consortium now collaborates with the Global Seed Vault initiative discussed below.

3. By 2008, the consortium included members in India, Australia, South Africa, Canada, the United States, and Britain (Lovell 2004; Frozen Ark Office 2006). The progress of the Frozen Ark has been slowed by protocols associated with animal rights law in Britain.

4. The science of genetics often contests theories of race and the suggestion that ethnic identity is biologically predetermined (see, for example, Lewontin and Levins 2007). However, as Pálsson and Rabinow (1999), Rabinow (1999), Simpson (2000), and Brodwin (2002) point out, the cultural construction of race has been refueled, in certain cases, by discussions that associate genetic identity with the boundaries of nation-states. Both Taussig (2004) and Pálsson (2007) present important examples of how the politics of national debates about new applications of genetics research are deeply embedded in particular cultural and historic contexts.

5. Donna Haraway (2003) notes the strange inversions of genetic essentialism between discourses on purebred pets and those on genetic diversity in endangered species. New cross-species cloning techniques suggest a continuing transformation of our kinship with nature.

6. Ian Wilmut, the genetic scientist responsible for the breakthrough, earned a membership in the Royal Society of Britain for his work.

7. On ethical debates and discussion of science related to cloning endangered and extinct species, see Amato et al. 2009, Corley-Smith and Brandhorst 1999, Comizzoli et al. 2000, Holt et al. 2004, Lavendel 2001, Lecard 2001, Loi, Barboni, et al. 2002, Loi, Galli, et al. 2007, Nash 2007, Ryder 2002, Ryder et al. 2000, Tallmon et al. 2004, and Vangelova 2003.

8. See Leader-Williams and Dublin 2000 for notes on charismatic large animal species, and Weeks 1999 and Vivanco 2002 for analyses of related environmental campaigns.

9. See Fletcher 2008a and 2008b for discussion of ancient DNA genomic research and cloning projects, particularly the attempt to resurrect the extinct thylacine (or Tasmanian tiger) in Australia.

10. Charis Thompson (2002, 2004) notes that interpretation of scientific debates over the "nature" of human/wildlife conflicts must attend to the political context and scalar orientation of the analysis. In the case of elephant conservation in Kenya, for example, the state's interest in ecotourism during the mid-1990s temporarily supported community-based conservation programs, working with the Maasai, over more globally oriented approaches to nature and wildlife protection. In Sardinia, government discourses have tended to support the latter. What Thompson (2002) calls multiple "competing philosophies of nature" are constantly in tension.

11. Already infamous for their historic resistance to subjugation by Roman, Spanish, and Piedmontese occupiers, Sardinian herding towns became notorious for banditry and blood feud following the unification of Italy. Criminologists in the late nineteenth century labeled this area a "delinquent zone" and measured people's heads to offer dubious theories about the biological roots of violence en-

demic to their "archaic," "primitive," and "atavistic" pastoral culture. See Heatherington 2010 for extensive discussion of parallels with environmental debates.

12. See Olson and Dinerstein 1998 for discussion of biodiversity "hotspots" used by IUCN, and Myers et al. 2000 for elaboration of the "Global 200" ecoregion approach adopted by WWF.

13. A distinction should be recognized between scientists' nuanced understandings of genetics and genomics and popular understandings of these fields. Genetic essentialism is embedded in certain public discourses despite recent research that supports sophisticated attention to cellular architecture and genetic expression, for example (see, for instance, Hogle 2003; Landecker 2007; C. Thompson 2007).

14. See, for example, Fraumene et al. 2003, as well as project descriptions on the website of the Institution of Population Genetics of Italy's National Research Council (http://www.cnr.it/istituti/DatiGenerali_eng.html?cds=038).

15. Wildlife conservation programs aim to maintain biological diversity at the ecosystem, population, species, and genetic levels (see, for example, Mangel et al. 1996). The term "genetic rescue" refers to efforts to increase the reproductive fitness of a population through the introduction of new alleles (that is, new genetic material). An expanding portfolio of genetic rescue techniques now includes the cryopreservation ("banking") of healthy cell samples (including germ cells and somatic cells representing the greatest possible intraspecies diversity) and a suite of new reproductive technologies, such as artificial insemination, in vitro fertilization, cytoplasmic sperm injection, nuclear transfer, and embryo transfer (see Comizzoli et al. 2000; Ptak et al. 2002). It should be stressed that interspecies "cloning" via nuclear transfer is not often a component of genetic rescue, but rather is still in experimental stages. According to Loi, Galli, et al. 2007, the problem of heterogeneous mitochondrial DNA in cloned mammal embryos continues to limit the number of surviving embryos that nuclear reprogramming can produce. Given the currently high inefficiency of this process, endangered species cloning programs are also limited by the availability of donor eggs (oocytes) and foster mothers for implantation, and by ethical concerns about the potential trauma to animal subjects.

16. Thom Van Dooren (2009a, 2009b) contends that seed banking has been used as a proxy for conservation, but that in fact it only guarantees that genetic resources remain available to agricultural researchers. The power contexts that condition these initiatives result in ex situ conservation that does little to serve farmers or the larger, messier project of stemming biodiversity loss.

17. Stephen Brush (2001) cautioned that the social implications of genetically modified organisms should not be conflated with environmental concerns. Modern crop science, he suggests, may well hold the key to improved food security for developing nations. For this reason, Brush has approved of seed banks but remained wary that they may be taken for a panacea for the erosion of crop genetic diversity, drawing attention and resources away from crucial in situ conservation measures. However, the embeddedness of new biotechnologies in a larger system

of global trade and intellectual property rights dominated by corporate interests, as well as existing economic inequalities, could further marginalize poor farmers and undermine sustainability (see P. Thompson 2007).

18. Some policy experts (see Elliot 2004; Renner 2006, 2007) have argued that conventional approaches to national and global security, relying on arms and military protection, have been ineffective and even counterproductive when applied to environmental security, and that we need to reevaluate the broader, very complex context of human security. Poverty alleviation, global food security, environmental health, just human access to ecosystem services, and peace are all considered vital elements of human security, which can only be achieved by putting new systems of collaboration and regulation into place across national borders.

REFERENCES

Acher, John
2008 Noah's Ark for Crop Seeds Opens in Arctic Norway. Reuters, February 26. Environmental News Network. http://www.enn.com/ecosystems/article/31808, accessed October 4, 2008.

Amato, George, Rob Desalle, Oliver A. Ryder, and Howard C. Rosenbaum, eds.
2009 Conservation Genetics in the Age of Genomics. New York: Columbia University Press.

Bowker, Geoffrey C.
2005 Time, Money and Biodiversity. *In* Global Assemblages: Technology, Politics and Ethics as Anthropological Problems. Aihwa Ong and Stephen J. Collier, eds. Pp. 107–123. Oxford: Blackwell.

Brodwin, Paul
2002 Genetics, Identity, and the Anthropology of Essentialism. Anthropological Quarterly 75(2):323–330.

Brush, Stephen B.
2001 Genetically Modified Organisms in Peasant Farming: Social Impact and Equity. Indiana Journal of Global and Legal Studies 9:135–162.
2004 Farmer's Bounty. New Haven: Yale University Press.

Brush, Stephen B., ed.
2000 Genes in the Field: On-Farm Preservation of Crop Biodiversity. Ottawa: IDRC. http://www.idrc.ca/openebooks/884-8/, accessed April 18, 2011.

Collier, Stephen J.
    2006 Global Assemblages. Theory, Culture and Society 23(2–3):399–401.

Comizzoli, Pierre, Pascal Mermillou, and Robert Maugei
    2000 Reproductive Biotechnologies for Endangered Mammalian Species.
    Reproduction Nutrition Development 40:493–504.

Convention on Biological Diversity (CBD)
    1992 United Nations Treaty Series no. 30619. Concluded at Rio de
    Janeiro, June 5, 1992. CBD Portal. http://www.biodiv.org/convention/
    default.shtml, accessed October 23, 2006.

Corley-Smith, Graham E., and Bruce P. Brandhorst
    1999 Preservation of Endangered Species and Populations: A Role for
    Genome Banking, Somatic Cell Cloning, and Androgenesis? Molecular
    Reproduction and Development 53(3):363–367.

Cuttelod, Annabelle, et al.
    2008 The Mediterranean: A Biodiversity Hotspot under Threat. *In*
    Wildlife in a Changing World: An Analysis of the 2008 IUCN Red List
    of Threatened Species. Jean-Christophe Vié, Craig Hilton-Taylor, and Si-
    mon N. Stuart, eds. Pp. 89–101. Gland, Switzerland: International Union
    for Conservation of Nature.

Dresser, Betsy
    2001 Nature's Nurturer. Interview by Constance Adler. Gambit Weekly,
    January 9. http://www.bestofneworleans.com/archives/2001/0109/covs
    .html, accessed October 23, 2006.

Elliot, Lorraine
    2004 The Global Politics of the Environment. 2nd ed. New York: New
    York University Press.

Fletcher, Amy
    2008a Bring 'Em Back Alive: Taming the Tasmanian Tiger Cloning Proj-
    ect. Technology in Society 30:194–201.
    2008b Mendel's Ark: Conservation Genetics and the Future of Extinc-
    tion. Review of Policy Research 25(6):598–607.

Fowler, Cary
    2008 The Svalbard Global Seed Vault: Securing the Future of Agriculture.
    The Global Crop Diversity Trust. http://www.croptrust.org, accessed
    October 8, 2008.

Franklin, Sarah
    2001 Sheepwatching. Anthropology Today 17(3):3–9.
    2003 Kinship, Genes and Cloning: Life after Dolly. *In* Genetic Nature/Cul-
    ture: Anthropology and Science beyond the Two-Culture Divide. Alan H.
    Goodman, Deborah Heath, and M. Susan Lindee, eds. Pp. 95–110. Berkeley:
    University of California Press.
    2007 Dolly Mixtures: The Remaking of Genealogy. Durham, NC: Duke
    University Press.

Fraumene, Cristina, Enrico Petretto, Andrea Angius, and Mario Pirastu
    2003 Striking Differentiation of Sub-populations within a Genetically Ho-
    mogeneous Isolate (Ogliastra) in Sardinia as Revealed by mtDNA Analysis.
    Human Genetics 114:1–10.

Frozen Ark Office
    2006 Saving the DNA and Viable Cells of the World's Endangered Animals
    for Their Future and Ours. Institute of Genetics, University of Nottingham,
    Nottingham, UK. http://www.frozenark.org, accessed October 23, 2006.

Fujimura, Joan
    2003 Future Imaginaries: Genome Scientists as Sociocultural Entrepreneurs.
    In Genetic Nature/Culture: Anthropology and Science beyond the Two-Cul-
    ture Divide. Alan H. Goodman, Deborah Heath, and M. Susan Lindee, eds.
    Pp. 176–199. Berkeley: University of California Press.

Haraway, Donna
    2003 Cloning Mutts, Saving Tigers: Ethical Emergents in Technocultural
    Dog Worlds. *In* Remaking Life & Death: Toward an Anthropology of the
    Biosciences. Sarah Franklin and Margaret Lock, eds. Pp. 293–328. Berkeley:
    University of California Press.

Hayden, Cori
    2003 When Nature Goes Public: The Making and Unmaking of Bioprospect-
    ing in Mexico. Princeton: Princeton University Press.

Heatherington, Tracey
    2010 Wild Sardinia: Indigeneity and the Global Dreamtimes of Environmen-
    talism. Seattle: University of Washington Press.

Heller, Chaia, and Arturo Escobar
    2003 From Pure Genes to GMOs: Transnationalized Gene Landscapes in
    Biodiversity and Transgenic Food Networks. *In* Genetic Nature/Culture: An-
    thropology and Science beyond the Two-Culture Divide. Alan H. Goodman,
    Deborah Heath, and M. Susan Lindee, eds. Pp.155–175. Berkeley: University
    of California Press.

Hogle, Linda
    2003 Life/Time Warranty: Rechargeable Cells and Extendable Lives. *In* Remaking Life & Death: Toward an Anthropology of the Biosciences. Sarah Franklin and Margaret Lock, eds. Pp. 61–96. Santa Fe: School of American Research Press.

Holt, William V., Amanda R Pickard, and Randall S. Prather
    2004 Wildlife Conservation and Reproductive Cloning. Reproduction 127:317–324.

Huxley, Aldous
    1989[1932]. Brave New World. New York: Harper & Row.

Intergovernmental Panel on Climate Change
    2007 Fourth Assessment Report (AR4). United Nations Environment Programme/ Intergovernmental Panel on Climate Change. http://www.ipcc.ch/, accessed January 2008.

Istituto Regionale Foreste Sardegna
    2006a Muflone: Ovis orientalis musimon. http://www.sardegnaforeste.it/ attivita/gestione_fauna, accessed October 25, 2006.
    2006b Risultati dei Censimenti del Muflone. http://www.sardegnaforeste.it/ attivita/gestione_fauna, accessed October 25, 2006.

Kay, Lily E.
    2000 Who Wrote the Book of Life? A History of the Genetic Code. San Francisco: Stanford University Press.

Landecker, Hannah
    2007 Culturing Life: How Cells Became Technologies. Cambridge, MA: Harvard University Press.

Lanza, Robert P.
    2002 Second Chances: An Interview with Robert P. Lanza. By Keith K. Howell. California Wild. 55(3):8–11. http://researcharchive.calacademy.org/ calwild/2002summer/stories/lanza.html, accessed May 9, 2011.

Lanza, Robert P., Betsy L. Dresser, and Philip Damiani
    2000 Cloning Noah's Ark. Scientific American 283(5):84–89.

Lavendel, Brian
    2001 Jurassic Ark. Animals, summer.

Leader-Williams, Nigel, and Holly T. Dublin
    2000 Charismatic Megafauna as "Flagship Species." *In* Priorities for the
    Conservation of Mammalian Diversity: Has the Panda Had Its Day? Abigail
    Entwistle and Nigel Dunstone, eds. Pp. 53–83. Cambridge: Cambridge Uni-
    versity Press.

Leadley, Paul, H. M. Pereira, R. Alkemade, J. F. Fernandez-Manjarrés, V. Proença,
J. P. W. Scharlemann, and M. J. Walpole
    2010 Biodiversity Scenarios: Projections of 21st Century Change in Ecosys-
    tem and Associated Ecosystem Services. CBD Technical Series 50. Secre-
    tariat of the Convention on Biological Diversity. http://www.cbd.int/doc/
    publications/cbd-ts-50-en.pdf, accessed July 3, 2010.

Lecard, Marc
    2001 Conservation Genetics in the Age of Genomics. Center for Biodiversity
    and Conservation Newsletter, American Museum of Natural History, fall–
    winter. http://congen.amnh.org, accessed October 23, 2006.

Lewontin, Richard and Richard Levins
    2007 Biology under the Influence: Dialectical Essays on Ecology, Agricul-
    ture, and Health. New York: Monthly Review Press.

Loi, Pasqualino, Barbara Barboni, and Grazyna Ptak
    2002 Cloning Advances and Challenges for Conservation. Trends in Bio-
    technology 20(6):233.

Loi, Pasqualino, Cesare Galli, and Grazyna Ptak
    2007 Cloning of Endangered Mammalian Species: Any Progress? Trends in
    Biotechnology 25(5):195–200.

Loi, Pasqualino, Grazyna Ptak, Barbara Barboni, Josef Fulka, Jr., Pietro Cappai,
and Michael Clinton
    2001 Genetic Rescue of an Endangered Mammal by Cross-Species Nuclear
    Transfer Using Post-mortem Somatic Cells. Nature Biotechnology 19:962–
    964.

Lovell, Jeremy
    2004 Modern Day Noahs Race to Build Wildlife Gene Bank. Planet Ark
    World Environment News, July 27.

Lowe, Celia
    2006 Wild Profusion: Biodiversity Conservation in an Indonesian Archi-
    pelago. Princeton: Princeton University Press.

Mangel, Marc, Lee M. Talbot, Gary K Meefe, et al.
    1996 Principles for the Conservation of Wild Living Resources. Ecological
    Applications 6(2):338–362.

Mellgren, Doug
2008 "Doomsday" Seed Vault Opens in Arctic: Frozen "Garden of Eden" Secures Biological Diversity for Future Generations. MSNBC, February 27, 2008. http://www.msnbc.msn.com/id/23352014/print/1/displaymode/1098/, accessed October 24, 2008.

Meyer, Stephen M.
2004 End of the Wild: The Extinction Crisis Is Over. We Lost. Boston Review, April–May. http://bostonreview.net/BR29.2/meyer.html, accessed October 25, 2006.

Myers, Norman, Russell A. Mittermeier, Cristina G. Mittermeier, Gustavo A. B. da Fonseca, and Jennifer Kent
2000 Biodiversity Hotspots for Conservation Priorities. Nature 403 (February 24):853–858.

Nash, Steven
2007 Millipedes and Moon Tigers: Science and Policy in the Age of Extinction. Charlottesville: University of Virginia Press.

Natural History Museum
2004 Frozen Ark Project Launches. News, July 27. http://www.nhm.ac.uk/about-us/news/2004/july/news_5295.html, accessed October 2006.

Nazarea, Virginia
2006 Cultural Memory and Biodiversity. Tucson: University of Arizona Press.

Olson, David M., and Eric Dinerstein
1998 The Global 200: A Representation Approach to Conserving the Earth's Most Biologically Valuable Ecoregions. Conservation Biology 12(3):502–515.

Ong, Aihwa, and Stephen J. Collier
2005 Global Assemblages, Anthropological Problems. In Global Assemblages: Technology, Politics and Ethics as Anthropological Problems. Aihwa Ong and Stephen J. Collier, eds. Pp. 3–21. London: Blackwell.

Pálsson, Gísli
2007 Anthropology and the New Genetics. Cambridge: Cambridge University Press.

Pálsson, Gísli, and Paul Rabinow
1999 Iceland: The Case of a National Human Genome Project. Anthropology Today 15(5):14–18.

Ptak, Grazyna, Michael Clinton, Barbara Barboni, Marco Muzzeddu,
Pietro Cappai, Marian Tischner, and Pasqualino Loi
    2002 Preservation of the Wild European Mouflon: The First Example of
    Genetic Management Using a Complete Program of Reproductive Biotech-
    nologies. Biology of Reproduction 66:796–801.

Rabinow, Paul
    1996 Artificiality and Enlightenment: From Sociobiology to Biosociality. *In*
    Essays on the Anthropology of Reason. Pp. 91–111. Princeton, NJ: Princeton
    University Press.
    1999 French DNA: Trouble in Purgatory. Chicago: University of Chicago
    Press.

Rajan, Kaushik Sunder
    2006 Biocapital: The Constitution of Postgenomic Life. Durham, NC: Duke
    University Press.

Renner, Michael
    2006 Introduction to the Concepts of Environmental Security and Environ-
    mental Conflict. Inventory of Environment Security Policies and Practices.
    Institute for Environmental Security. http://www.envirosecurity.org/ges/
    inventory, accessed July 3, 2010.
    2007 Worldwatch Perspective: Security Council Discussion of Climate
    Change Raises Concerns about "Securitization" of Environment. World-
    watch Institute Online Features, Eye on Earth, April 30. http://www.worldwatch
    .org/node/5049, accessed July 3, 2010.

Roug, Louise
    2007 The Svalbard Seed Vault atop the World. LA Times online, October 12.
    http://articles.latimes.com/2007/oct/12/world/fg-vault12, accessed October
    11, 2008.

Ryder, Oliver A.
    2002 Cloning Advances and Challenges for Conservation. Trends in Bio-
    technology 20(6):231–232.

Ryder, Oliver A., Anne McLaren, Sydney Brenner, Ya-Ping Zhang, and
Kurt Benirschke
    2000 Ecology: DNA Banks for Endangered Animal Species. Science
    288(5464):275–277.

Sample, Ian
    2004 Frozen Ark to Save Rare Species. Guardian, July 27. http://www.guardian
    .co.uk/science/2004/jul/27/biodiversity.environment, accessed October 5,
    2008.

Sanne, Dyveke
2008 Interview by Miranda F. Mellis. The Believer 6(9):34–36. http://web
.me.com/dyvekesanne/www.dyvekesanne.com/svalbard_global_seed_vault
.htm, accessed July 4, 2010.

Shears, Richard
2008 Extinct Tasmanian Tiger Could Roar Back into Life after DNA Is
Implanted into a Mouse. Mail Online, May 20. http://www.dailymail.co.uk/
sciencetech/article-1020675/Extinct-Tasmanian-tiger-roar-life-DNA-
implanted-mouse.html, accessed October 5, 2008.

Simpson, Bob
2000 Imagined Genetic Communities: Ethnicity and Essentialism in the
Twenty-First Century. Anthropology Today 16(3):3–6.

Tallmon, David A., Gordon Luikart, and Robin S. Waples
2004 The Alluring Simplicity and Complex Reality of Genetic Rescue.
Trends in Ecology and Evolution 19:489–496.

Taussig, Karen-Sue
2004 Bovine Abominations: Genetic Culture and Politics in the Netherlands.
Cultural Anthropology 19(3):305–336.

Thompson, Charis
2002 When Elephants Stand for Competing Philosophies of Nature: Am-
boseli National Park. *In* Complexities: Social Studies of Knowledge Practices. John Law and Annemarie Mol, eds. Pp. 166–190. Durham, NC: Duke
University Press.
2004 Co-producing Cites and the African Elephant. *In* States of Knowledge:
The Co-production of Science and the Social Order. Sheila Jasanoff, ed. Pp.
67–86. London: Routledge.
2007 Making Parents: The Ontological Choreography of Reproductive Technologies. Cambridge, MA: MIT Press.

Thompson, Paul B.
2007 Food Biotechnology in Ethical Perspective. 2nd ed. Dordrecht, Netherlands: Springer.

Van Dooren, Thom
2009a Banking Seed: Use and Information in the Conservation of Agricultural Diversity. Science as Culture 18(4):373–395.
2009b Genetic Conservation in a Climate of Loss: Thinking with Val Plum-
wood. Australian Humanities Review 46. http://www.australianhumanities
review.org/archive/Issue-May-2009/vandooren.html, accessed July 13, 2010.

Vangelova, Luba
    2003 True or False? Extinction Is Forever. Smithsonian Magazine, June.
    http://www.smithsonianmag.com/science-nature/True_or_False_Extinction
    _Is_Forever.html, accessed October 21, 2006.

ViaGen
    N.d. Our Services: Gene Banking. ViaGen, The Cloning Company. http://
    www.viagen.com/en/our-services/gene-banking, accessed October 5, 2008.

Vivanco, Luis. A.
    2002 Seeing Green: Knowing and Saving the Environment on Film. Ameri-
    can Anthropologist 104(4):1195–1204.

Weeks, Priscilla
    1999 Cyber-activism: World Wildlife Fund's Campaign to Save the Tiger.
    Culture & Agriculture 21(3):19–30.

3. TOTEM AND TABOO RECONSIDERED

ENDANGERED SPECIES AND MORAL PRACTICE IN MADAGASCAR

Genese Marie Sodikoff

It is midday, hot, and we are squeezed haunch to haunch on wooden benches in the open flatbed pickup. Our bush taxi nears the bustling roadside village of Sandrakatsy, lying along the Mananara River of northeast Madagascar. We spy something in the ditch beside the road: a white owl bound to a cross made of tree branches, wings extended and downy head slung forward on its breast. I wonder what it means, but since no one in the taxi speaks, I dare not ask.

Later that day, I interrogate Navony, an authority on traditional matters, about the crucified owl. She is a woman in her mid-sixties and the wife of the *tangalamena* of Varary. This is a village that sits a few miles off-road near the boundary of a temperate rain forest that forms the nucleus of the Mananara-Nord Biosphere Reserve. (I conducted research in this region on conservation and low-wage labor for fourteen months between 2000 and 2002 [Sodikoff 2009].) "Tangalamena" (literally "red baton") is the title given to elder spiritual leaders of Betsimisaraka villages. Betsimisaraka form the dominant ethnic population of eastern Madagascar. The tangalamena of Varary, Navony's husband, presides over rituals, keeps the oral history of the village in his memory, and communicates with dead ancestors. He and Navony have a son who works for the conservation project overseeing the Mananara-Nord Biosphere Reserve.

Navony explains to me that people fear and hate white owls. "*Vorondolo* [spirit birds]," she calls them, "the playthings of sorcerers." Sorcerers (*mpamosavy*) in Madagascar are known to dabble in black magic, and it always seemed

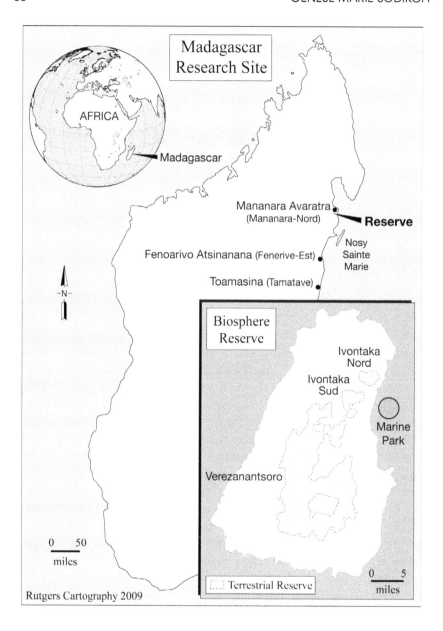

Figure 3.1. Map of the Mananara-Nord Biosphere Reserve. The inset shows the names of several stands of primary rain forest that form the national park of the reserve. Created by Rutgers Cartography Services, 2009.

to me that any villager who had the good fortune to reach ripe old age was inevitably accused of being a sorcerer. She adds that it is taboo (*fady* or *faly*) around those parts to eat white owls, but it is good to kill them because doing so foils their evil commissions.

Fady, as any scholar of Madagascar will attest, form an integral part of Malagasy social life, regardless of class, locality, or ethnic identification (Van Gennep 1904; Frazer 2000; Ruud 1959). Fady are prohibitions established by dead ancestors who wield great power over the living, and whose wishes are upheld by elders, the next in line to ancestorhood. The proscription of an object—be it plant, animal, landscape feature, day of the week, cigarette, word, gesture, transaction, or whatever—ascribes sacredness and danger to the object. Animal fady, a category of taboo, impose bodily and linguistic constraints; they prohibit the killing of the animal and sometimes require that one avert one's gaze from it, use euphemisms for it, or avoid particular land features or resources tied to it (Bodin et al. 2006).

Until recent times, according to Navony, people had obeyed the fady against eating the white owl. Nowadays, however, "some young people dare to eat it." Navony feared that the transgression not only angered dead ancestors but also, in the grander scheme of things, threatened the historical identity of her people. In the early 2000s, the town and surrounding countryside of Mananara-Nord was already undergoing a quick-paced social transformation prompted in large part by the boom in international prices for cloves and vanilla. People were streaming into the region to make their fortunes in cash cropping. The Mananara-Nord Biosphere Reserve had also become fully operational as a tourist destination and scientific research site, which meant that people's actions within the reserve would at times be monitored and penalized.[1] Betsimisaraka villagers experienced these changes with unease and sometimes fear. The influx of outsiders tended to increase nonviolent crimes, like the theft of vanilla plants and bags of harvested cloves, as well as raucous behavior in town. Elders did not like seeing ancestral customs, such as fady obedience, disrespected.

In this chapter, I examine the meanings of animal taboos and taboo animals in a context of expansive conservation interventions since the late 1980s and rapid social change in eastern Madagascar (see Kull 1996). Animal taboos and taboo animals are objects around which two extinction events converge: the one cultural (the loss of ancestral custom), the other biological (the loss of endemic species). Both events instill a sense of regret and risk, though for different reasons and with respect to different social groups. Betsimisaraka elders fear ancestral retribution for acts of transgression, and conservation representatives at higher administrative levels worry about biodiversity loss and know that the abandonment of certain animal fady by rural Malagasy only exacerbates the problem. Despite the mutual consternation, these groups do not find common cause.

By comparing how Malagasy and Westerners view wild species and enforce taboos around them, I trace here how Malagasy animal fady have been invested with new meanings as the moral code of conservation asserts itself. In protected areas like the Mananara-Nord Biosphere Reserve, the explanation to residents of rules spelled out in the Convention on Biological Diversity and the Convention on the International Trade in Endangered Species of Wild Flora and Fauna (CITES) cannot but lend new meaning to animal fady and acts of fady transgression. If an animal happens to be doubly proscribed, at once fady and on the International Union for Conservation of Nature's (IUCN) Red List of Threatened Species, then, from a Malagasy perspective, fady transgression may become an act of resistance to foreign authority, an expression of cultural self-determination, and therefore an instance of moral practice. In what follows, I investigate the meanings of the taboo animal to glean how Betsimisaraka residents of a protected area speak to power through animals.

Taboo and Totem in Madagascar

Betsimisaraka literally means "the many who do not sunder." Several other names of ethnic groups in Madagascar also incorporate the negative *tsy* (or *tsi*). Negative action, such as "not sundering," is the essence of taboo, a self-imposed stricture that guides moral practice, particularly in contexts where the state has minimal oversight over people and therefore does not legislate people's actions (Valeri 2000). Obedience to fady thus contributes to the formation of moral personhood, as well as individual, family, and group identity. Michael Lambek (1992) argues that in Madagascar, fady set the parameters of moral practice by creating cognitive, spatial, temporal, bodily, and societal boundaries. Proscribed actions distinguish selves and groups from one another. An individual's transgression of an ancestral taboo can cause a sense of moral disorder among the social group to which he or she belongs (Kus and Raharijaona 2000).

Many North Americans' reports on and accounts of travel in Madagascar in the past two decades include descriptions of the abandonment of cultural taboos that have protected lemurs from extinction. Lemurs are a family of species which rank among the world's most endangered (see Godfrey and Rasoazanabary, this volume). In 2005, the IUCN listed 16 percent of all lemur species as Critically Endangered, 23 percent as Endangered, 25 percent as Vulnerable, and only 8 percent as Least Concern (28 percent were "Data Deficient"; Mittermeier et al. 2006). They are Madagascar's flagship species family. The categorization of species from most to least endangered constructs a hierarchy of species value. At the top tiers are the easily observable, charismatic fauna.

The following passage from Alex Shoumatoff's blog *Dispatches from the Vanishing World* is typical. Shoumatoff relates his 1988 visit to the Special Indri Reserve of Andasibe, Madagascar, where for decades Betsimisaraka residents of the region had upheld a fady against killing the indri, the largest of Madagascar's extant primates. He had asked a passerby about the consumption of lemurs by residents of the area:

> "They know indri is an ancestor, but they eat him anyway," Maurice explained.
>
> "Why? Isn't it fady?" I asked. "In the old days, it was very fady," the spokesman for the woodsmen went on. "As fady as marrying your sister."
>
> "So why do you eat indri?" I repeated.
>
> "Because they are fat and because today there is a shortage of protein. We have to eat. A kilo of beef costs a thousand francs in the market at Andasibe. A lemur may weigh eight kilos, and we can sell it for fifteen hundred francs a kilo."
>
> "Don't you get punished with sickness for violating the fady?"
>
> "Others get sick, even die, because of their belief," the spokesman said.
>
> "These men don't have the belief deeply," Maurice explained. (Shoumatoff 1988)

The Malagasy "spokesman" for the woodsmen offers a biological reason for the transgression of the fady—the human need for protein—as well as a cultural-ecological reason, the transformation of belief in the regional landscape brought on by the in-migration of people from other regions. It is also noteworthy that the spokesman fearlessly confesses to a foreigner that the crime of indri killing is commonplace there.

The spokesman for the woodsmen also offers an origin story for the indri fady: "indri is an ancestor." One hears variants of this story around the island. For anthropologists, this is an artifact of totemic thought (Shaw 1896; Van Gennep 1904). Early anthropologists held that the totem, typically an emblematic animal of the social group, possessed the status of the group's apical ancestor, thus drawing a line of descent from nonhuman animal to human. The negative actions directed at the taboo object serve the added functions of differentiating groups from one another, and naturalizing social classifications to establish, for example, the rules of marriage and descent, as Claude Lévi-Strauss argued (Mullin 1999:208).

Figure 3.2. A species of mouse lemur. Photograph by Genese Marie Sodikoff, 2001.

Victorian and early post-Victorian anthropologists, preoccupied with the origins of taboos, speculated that taboos were survivals of clan, village, or family totems (Shapiro 1991). For ethnologists of the late nineteenth and early twentieth centuries, Madagascar was a good place to study the origins and meaning of taboo. Sailing from the Malay-Polynesian archipelagos approximately two thousand years ago, the island's first settlers brought the "taboo idiom" from Austronesia to Madagascar (Steiner 1956:147). Early ethnographers found evidence in Malagasy society of an imagined bond of descent between people and the endemic primates. In theorizing totemism, these ethnologists perhaps forced the matter of literal belief over folklore. G. A. Shaw (1896:201–

203) wrote that Betsimisaraka people conceived of the aye-aye (*Cheiromys madagascariensis*) as "the embodiment of their forefathers, and hence will not touch it, much less do it an injury. It is said that when one is discovered dead in the forest, these people make a tomb for it and bury it with all the forms of a funeral. They think that if they attempt to entrap it, they will surely die in consequence."

Arnold van Gennep (1904:214–294) similarly collected Malagasy tales of origins of animal fady, particularly those tied to endemic primates. An animal may be regarded as the generator or sibling of an individual, a reincarnated human being, or a metamorphosed ancestor. Jørgen Ruud records a Tanala story of the babakoto ("grandfather"), a name that appears to apply to both *Lichanotus brevicaudautus,* another large lemur species, and *Indri indri.* A man in the forest is trying to tap rubber high up in a tree and slips. He nearly crashes to the ground, but a babakoto comes to the man's rescue and delivers him safely. In eternal gratitude to the babakoto, the man and his people vow that from then on, it is fady to injure or kill the animal (Ruud 1959:99). Origin stories of animal fady often involve events in which an animal helped or harmed one's ancestors, so avoiding it signifies people's eternal gratitude or wariness (Van Gennep 1904). Not only the prosimians but also other species were included in the totemic kinship. James G. Frazer (1922:519) reiterates the literalist interpretation of the human/animal genealogical bond when he notes that various Malagasy ethnic groups "believe themselves to be descended from crocodiles, and accordingly they view the scaly reptile as, to all intents and purposes, a man and a brother."

Recent anthropological analyses of taboo are less concerned with deciphering origins than in examining what role the institution of taboo plays in contemporary social life. Fady in Madagascar simultaneously invent and sustain a code of moral conduct for the individual or group. A commonly held fady can fortify group solidarity (Graeber 1995:265). Abiding by any fady thus defines the self (and group) in terms of what it does not do (Valeri 2000). In this way, taboo obedience shapes a sense of moral selfhood, while transgression threatens one's moral integrity (Lambek 1992).

While all kinds of fady are constantly being established, abolished, and transformed within Malagasy societies, Malagasy people have also used fady as a means of delineating the boundaries of difference between Malagasy and foreigners, particularly French colonialists, and of defending those boundaries by resisting domination and cooptation. Fady have served to mobilize resistance to colonial power (Feeley-Harnik 1984). Fady days, or days of the week on which one may not work, have served as a means of resistance to outsiders' demands on Malagasy labor (Jarosz 1994). They have served to protect territory, as in the case of the Mahafale, who proclaimed it fady to let white foreigners enter their territory; Europeans (*vazaha*) were "considered so strange that their other name is *biby,* animal" (Kaufmann 1999:138). So in addition to the moral strictures they establish to distinguish difference and to contain danger—thereby constructing that very danger (Douglas 2004)—fady have also signified resistance to the encroachment and authority of outsiders.

The Fady Strategy of Conservation

Scientists interpret the rise in the abandonment of taboos concerning endemic wild animals, such as lemurs, sea turtles, and certain species of fish and fowl, to be driven by people's need for new sources of wild protein as preferred species die out due to habitat erosion (Jones et al. 2008; Rabearivony et al. 2008). North American, European, and Malagasy conservation representatives have tried to reinforce the intrinsically "conservationist" fady in Madagascar, hoping that if conservation is connected to the taboo idiom, rural people will be more inclined to obey conservation rules and to support the conservation effort, such as by adopting agricultural techniques that do not depend on cutting down and burning forest vegetation. Eva Keller (2009) argues that in their attempt to think about Malagasy culture and how cultural beliefs might hinder or serve conservation, many conservation representatives reduce the complexity of Malagasy culture to taboo. They believe that if conservation rules are presented in terms of fady, then Malagasy people will buy into the conservation program rather than resist it.

The simplest approach by conservation representatives has been to substitute the word "fady" for "prohibited"—that is, *rarana* in Malagasy or *interdit* in French—on signs around protected areas. In Ankarana, northern Madagascar, for example, an official sign at the boundary of a special reserve reads *Ala fady* to deter trespassers (Walsh 2005:654). "Ala fady," meaning "sacred or taboo forest," is a widespread "habitat taboo" in Madagascar. It refers to primary forest that shelters family tombs, and ethnic groups in forest areas respect the prohibition against razing and burning such forest in order to cultivate crops (Colding and Folke 2001:590). Conservation representatives' use of the phrase "Ala fady" at the boundary of a nature reserve exploits people's fear of ancestral retribution.

Malagasy people who live near or within protected areas have a greater number of "taboos" tied to local species because they are subjected to conservation regulations more directly than people who live on the outskirts. They are perhaps more inclined to perceive certain wild animals as agents (or wards) of "neocolonial" authority. The strategy of coopting fady to benefit the conservation program has been generally ineffective. Between 2000 and 2002, however, residents of the Mananara-Nord prefecture appeared to have a keen awareness of conservation representatives' interest in fady that pertained to wild species and landscapes. In my view, this awareness suggested how the cultural significance of animal fady was shaped by the multinational conservation effort and local resistance to it. Fady transgression must be interpreted in light of the imposed legal taboos against killing animals which appear on IUCN's Red List of Threatened Species.

The New Totemism

Taboo is conceptualized as a dynamic moral stricture and political practice (a repudiation of outsider authority) not merely by marginal societies of the global South but also in secular, industrialized societies. While legal proscriptions against killing endangered species lack the supernatural power of fady, where transgression is thought to be monitored and punished by dead ancestors, such juridical taboos are also moral strictures that, from the Malagasy perspective, define the moral selfhood of Westerners.

During intermittent periods of fieldwork in Madagascar between 1994 and 2002, I often heard rural Malagasy people complain that "vazaha [foreigners] care more about lemurs than people!" To many Malagasy, Westerners' apparently greater concern for the well-being of fauna has historical continuity. Conservation in Madagascar was codified during the French colonial regime (1896–1960); by the 1920s colonial powers were collaborating to protect the natural heritage of each distinct colony. The reinvigoration of conservation efforts in the 1980s, after a postindependence interim, has entailed a campaign that promotes the conservation of Madagascar's "natural heritage." Arguably, the natural heritage of unique tropical regions defines these places in the Western imagination more than their cultural heritage, perhaps because cultural beliefs and practices are often vilified as causes of species extinction.

As Peter Whiteley elaborates in the epilogue to this volume, the charismatic endangered megafauna that are key symbols of the global conservation effort also represent the menagerie of a "new totemism." Such species are sacralized through juridical taboos against killing, hunting, or consuming them, and they become components of a national natural heritage and national identity. The new totems provide national origin stories by forging a genetic tie to the territory through endemic species (Comaroff and Comaroff 2001). Flagship animal species, such as Malagasy lemurs, the giant panda and river dolphin (baiji) of China, the Congolese gorilla, and the Galápagos sea turtle, help stir interest in and raise funds for transnational conservation activities.

The association of nature and nation has a deep history. State-builders have long recruited plants, animals, and landscapes into myths of national autochthony. Ruling classes have grafted nonhuman species onto the origin myths of nations as a means to naturalize and eternalize a social order. The idea of origins fixes the composition of ecosystems and human populations in time, as well as motivating attempts to protect the claims and identities of certain groups against others. Wilderness is imagined as original and static, antecedent to culture.

In the United States, anthropomorphized, endangered animals in children's television foster a sense of moral obligation and protectiveness toward animals. For example, in 2002 I was in Madagascar reading an issue of one of Madagascar's daily newspapers, *L'Express*. A story about an organization called "Aiza Biby" caught my eye. "Aiza Biby" ("Where Are the Animals") was a New York–based association founded by conservation biologists with research ties to Madagascar. A photo in this Malagasy newspaper showed a person dressed as a white-collared lemur entertaining a group of children at the old army fort in Brooklyn, New York. "The families of the New York firefighters [*pompiers new-yorkais*] now know the 'varika' thanks to an initiative of the association Aiza Biby," the text reads (L'Express de Madagascar 2002). Now at this point in time, four months after September 11, 2001, Malagasy and expatriate readers would have mentally connected "*pompiers new-yorkais*" to the terrorist strike of September 11. Readers might interpret the visit of a human in a varika costume as a diplomatic gesture of sympathy from Madagascar to the United States, specifically to American firefighters' children. In Western cultures of childhood and practices of consolation, animals warm the chill of absence. Aiza Biby's primary mission at the time, however, was to educate the public about the causes of, and solutions to, species endangerment in Madagascar. The varika character alerts the bereft families of New York firefighters to another calamity: the imminent extinction of lemur species. Supporting the conservation effort constitutes moral practice.

Lemurs become the missionaries of their own salvation. As scientists discover new species of lemur in Madagascar, they bestow moralistic names. The taxonomic name establishes a relationship of guardianship and paternity. The practice has a long history in Madagascar, where numerous species carry the names of European naturalists who explored the island (see Dorr 1997). The name of *Microcebus lehilahytsara,* the mouse lemur "good man," pays tribute to the North American lemur expert Stephen Goodman. *Avahi cleesei,* a species of miniature woolly lemur, was named after British comedian John Cleese in honor of his lemur advocacy in film (Wildmadagascar.org 2005). *Mirza zaza* is dedicated "to Madagascar's children, to remind them of their responsibility for preserving the island's unique biodiversity for future generations," according to a press release from Chicago's Field Museum (Chamberlain 2005); *zaza* is Malagasy for "child." All of these are members of *Cheirogaleidae*, the family of mouse and dwarf lemurs, and the discoveries of these new species inspired a North American primate biologist to proclaim that "this tiny creature has become a huge ambassador for all things wild in Madagascar" (Environment News Service 2006). As an ambassador of wild Madagascar, the varika stirs up interest in and concern for Madagascar's extinction problem caused by human actions.

Fady and Unspoken Resistance

The Mananara-Nord Biosphere Reserve in 2002 encompassed a terrestrial and marine park as well as 250 Betsimisaraka hamlets occupied by peasants and fishermen. Residents were continually reminded of conservation rules, including the ban on killing lemurs, the ban on killing any animals within the boundaries of the terrestrial park, and the strict rules concerning fishing in the marine park. In the following account, I describe an occasion on which animal fady were deployed as resistance to the will of the conservation authority. Ironically, this resistance came from within the conservation project itself, through an employee of the Biosphere Reserve who also identified as a local villager and adherent of ancestral ways. The account demonstrates the ways in which faunal species and taboos have in recent times been resignified within the politicized context of protected areas, where outsiders' moral codes collide with "indigenous" ones.

The Rats of Nosy Antafaŋa

The marine park of the Mananara-Nord Biosphere, a plot of sea on the lower lip of Antongil Bay, includes a coral reef and three islets (*nosy*), each named for its predominant feature: Germania Tree Island (Nosy Antafaŋa); Bird Island (Nosy Vorona), for the flocks of Saunders' terns that fly above it); and Rangontsy Island (Nosy Drangontsy), girded by mangroves and named after three brothers whose bones lie boxed there in the shade of a granite outcrop. Named for and bearing trees, birds, and brothers (the Rangontsy brothers are dead ancestors now)—plants, animals, and humans—the islets represented members of a moral community that had found itself in the throes of ecological adjustment measures since 1989, when the Biosphere Reserve project entered the scene.

On Nosy Antafaŋa, the largest of the islets, the rats are profuse and bold, so much so that belongings must be slung over beams or stored in metal chests weighted shut with stones to keep them from being ransacked—even in broad daylight, even as people watch. Nosy Antafaŋa was uninhabited by humans in the early 2000s except for a man named Ali, whose austere, one-room house was overrun by rats. The rats did not bother him. In fact, Ali respected them. He was a 56-year-old employee of the Biosphere Reserve project, and he had been working for the conservation project since 1991. He patrolled the Biosphere's marine park in his motorized outrigger canoe, confirming that the men spearing octopus or netting fish were authorized to fish in it.

Ali described himself as "severe" with the fishermen and unpopular in the village of Sahasoa, which lay on the mainland opposite the islets. Andrew Walsh argues that in Madagascar observing a taboo elevates one's moral status in society only if it means renouncing something one might otherwise be able to do (Walsh 2002, 2006:6). Responsible and ethical practice is situated relative to emergent social relationships and economies. Ali, as a newcomer to the village and as a representative of the Mananara-Nord Biosphere Reserve, might have rejected the local fady. But despite how Ali described himself, Sahasoa residents liked him because he chose to obey the local fady concerning five kinds of animals on the islets: the skink (a type of large lizard), the tern, the rat, the zebu, and the wild boar. Of the five, only skinks, terns, and rats occur there naturally, in the "wild." Cattle and boar, in contrast, appear only as meat brought over by visitors, which was fady. It was also fady to kill, eat the meat of, injure, or utter the real names of these five animals while on the islets. Euphemisms had to be used instead.

Common name		Euphemism
androngo (skink)	→	*sahidilahy* (male witness?)
sikôza (tern)	→	*tsy mirirana* (doesn't curve, doesn't lean)
voalavo (rat)	→	*tsy mamaky* (doesn't traverse, doesn't read)
aomby (zebu)	→	*lava vava* (long mouth)
lambo (boar)	→	*antaniloha* (in the original country)

No one could tell me what the euphemisms meant or anything about their origin. The strangeness of the euphemisms, the nonsense of them, served to distance the speaker from danger.

A possible origin of the euphemisms is suggested by James Sibree, who describes the Malagasy custom of designating a euphemism for animal names shared with chiefs or royalty:

> The names of the chiefs almost always contain some word which is in common use by the people. In such a case, however, the ordinary word by which such thing or action has hitherto been known must be changed for another,

which henceforth takes its place in daily speech. Thus, when the Princess Rabodo became queen in 1863, at the decease of Radama II, she took a new name, Rasoherina . . . Now soherina is the word for chrysalis, especially for that of the silkworm moth; but having been dignified by being chosen as the royal name, it became sacred (fady) and must no longer be employed for common use; and the chrysalis thenceforth was termed zana-dandy, "off-spring of silk." So again, if a chief had or took the name of an animal, say of the dog (amboa), and was known as Ramboa, the animal would be henceforth called by another name, probably a descriptive one, such as fandroaka, i.e., "the driver away," or famovo, "the barker." (Sibree 1892:226–227)

It is therefore possible that the Rangontsy brothers of Sahasoa had a chiefly status in life and had taken on several of the animal names in question, which would prompt a change in their vernacular names.

Ali enjoyed a good reputation in Sahasoa because he observed the taboos of Nosy Antafaŋa, respecting the rats and, by association, the Rangontsy ancestors. Connected to the negative actions surrounding the five animals was the positive action of paying tribute to the Rangontsy brothers when visiting their eponymous islet. On a day in February 2002, Ali invited me on a day-long tour of Nosy Drangontsy. He warned me ahead of time to bring some small change, but I could not find a single coin in Sahasoa village before our departure. No one in the village could break a bill for me. Ali told me not to worry about it. At Nosy Antafaŋa, Ali beached the pirogue, and from there we waded through the shallow tide and between the umber roots of mangroves to Nosy Drangontsy. Setting foot on Nosy Drangontsy, we headed toward its center, where the tombs (*fasaŋa*) of the brothers lay. The *fasaŋa* consisted of three boxes of bones nestled under an outcropping and marked by a wooden sign. We stood in front of the tomb, and Ali began to make a formal speech, introducing me to the ancestors. When he finished, he turned away. Then, unexpectedly, he spun back to face the tombs again and yelled out, "She will learn! She will learn!" At that moment, he revealed how anxious he was about my failure to offer any coins.

At the day's end, we prepared to boat back to the mainland. A few minutes out to sea and the motor died. Ali swore to himself, shaking his head. I could see he was uneasy. The wind was picking up, and waves lapped the sides of the pirogue. We were too far from shore to row home against the tide. I worried about Ali's disappointment in me for not having paid tribute to the Rangontsy brothers, and for accidentally having uttered the word "lambo" for boar instead of its euphemism, "antaniloha," the day before on Nosy Antafaŋa. I also worried I had shown disrespect, having blurted "Oh! *Tsy mamaky*!" when a large rat lunged onto the table at his house, because even though I had used the correct euphemism, it was apparently wrong to address the rats directly. Fortunately, the boat's motor restarted after ten minutes or so, and we arrived safely on shore. Later in the evening, over beers, Ali confessed he had thought the ancestors killed the boat's motor.

He proceeded to tell me about another occasion in the late 1990s when the motor died, leaving the boat adrift. It was the period when administrators of the Biosphere project had attempted to eradicate all the nonnative species on Nosy Antafaŋa. Biosphere workers had to uproot all the lemon and breadfruit trees, banana plants, and coconut palms. And since it threatens the survival of endemic and native species, conservation scientists wanted to exterminate *Rattus rattus*. Ali claimed that the Biosphere bosses had promised him and Sahasoa residents they would leave the rats alone, but in fact they hired a specialist to poison them. The animals were baited with wax blocks saturated with anticoagulant poison (Cooke et al. 2003:207). "The Biosphere told a big lie back then," said Ali; "many many rats died." Killing the rats risked moral disorder. The sea could become choppy and unfishable, and defunct motors could strand men at sea. As proof, the motor of the Biosphere project's canoe died for good, and the project had to buy a new one.

By 2000, Ali and the villagers had nevertheless survived the anger of the rats, and the species had rebounded spectacularly on Nosy Antafaŋa. The Biosphere Reserve's conservation project had abandoned its rat eradication plans in order to improve relations with the village. The conservation authorities recognized the value of sacrificing the endemic purity of an islet to gain the Sahasoa community's support of the conservation effort in the marine park. Ali became the spokesperson for Sahasoa fishermen and villagers vis-à-vis the Biosphere Reserve bosses. He not only remained vigilantly observant of the animal fady, he had become vigilant about the transgressions of the Biosphere administrators. He argued that if fishermen had to obey conservation rules, then the conservation project ought to respect local fady.

The animal taboos of Sahasoa were used by residents to localize and contain the threat of the conservation project, which in turn threatened to disrupt the social order. Adherence to the islet taboos paid tribute to the Rangontsy patriarchs, nurturing the group identity of Sahasoa. But knotted into the meaning of the local fady now was resistance to the kinds of transformations sought by the conservation project. Situating their ethics and moral practice with respect to foreign interventions they had little power to prevent (see Walsh 2005), Sahasoa residents made concessions to external authority but rallied around custom. They also demanded concessions from the conservation project staff, as if to challenge the project's claim to desire local participation and collaboration. They asserted that a species considered by conservationists to be an exotic, invasive rodent, and therefore utterly devalued, was indeed valuable.

Conclusion

Public reports about Madagascar's spiraling environmental crisis often discuss the abandonment of animal fady in Madagascar as though it is directly the result of the slash-and-burn agriculture practiced by Betsimisaraka subsistence farmers, which has eroded the rain forest significantly and has long been considered by scientists to be the main cause of deforestation. The logic is that deforestation depletes faunal species, so people have fewer animals to hunt. They abandon or abolish fady that become detrimental to survival. But taboo abandonment is not analyzed within a broader political context of capitalist expansion and foreign conservation intervention. Red List taboos intrude on the fady system and work to replace the moral stricture of customary taboos with conservation legislation and its quasi-totemic hierarchy of species value. The signification of each taboo system shapes the other, calling into question what defines moral personhood. Conservation legislation is informed not solely by the objective science of biodiversity loss but also by an ideology of animals and animal rights. Many Western environmentalists deem animals to possess moral considerability, or intrinsic value, and therefore the right to exist as a species. Conservation rules are shaped by a culturally specific morality and a strategic totemism. Endangered flagship species represent and inform national identities, drum up support for international conservation efforts, and disseminate a culture of animal love familiar to Western societies.

In Madagascar's protected areas, an individual's conscious transgression of a fady against killing lemurs, sea turtles, birds, or some other endangered species may not only offend the ancestors (if one ascribes to that belief) but also signify resistance to outsider authority. If fady transgression is resignified as an act of resistance, it has the potential to defuse ancestral wrath because it seeks to conserve cultural identity and autonomy. This possibility presented itself one day in December 2001 when a Betsimisaraka political appointee in the town of Mananara-Nord flagrantly shot four lemurs in the core of the Biosphere Reserve while being videotaped by an accomplice. He then broadcast his hunt and the four kills on the local television station when the minister of the environment was visiting the town. The hunter-politician knew that his rural constituency would applaud his defiance of the conservation effort, represented locally by the Biosphere Reserve. Even though he himself did not abide by any lemur fady, he demonstrated publicly how killing endangered species could be a powerful act of political resistance and assertion of cultural self-determination (despite the fact that virtually no one hunts with guns there).

At the same time, an individual who rigidly upholds a fady that protects invasive exotic species, such as skinks, rats, and feral boars, also does so with a conscious knowledge of how such an observance subverts conservation plans designed to restore a biosphere to a time when it was inhabited only by "endemic" species. To obey or transgress one kind of taboo or another in Madagascar involves a mental calculus of both personal risk and cultural survival. Expatriate and national conservation representatives in turn have learned by trial and error what one must not do if one hopes to open the doors of communication with Betsimisaraka residents.

ACKNOWLEDGMENTS

I would like to thank people who have given me helpful suggestions and comments on different versions of this paper, including Jill Constantino, Gillian Feeley-Harnik, David Graeber, Michael Hathaway, David Hughes, Molly Mullin, and Harriet Ritvo.

NOTES

1. More recently, conservation efforts in the Antongil Bay region have been derailed by illegal logging of rosewood and ebony in the island's rain forest parks. This began after the ousting of President Marc Ravalomanana by *coup d'état* in March 2009. The coup led to widespread political unrest. Protected areas along the east coast, including the Mananara-Nord Biosphere Reserve, were beset by Chinese and Malagasy "timber barons" who hired and armed gangs of men to illegally fell rosewood and ebony and ship it to China, Europe, and the United States for large profits (Gerety 2009). These work gangs have set up encampments in the forest, and they have been known to intimidate villagers and collude with local officials.

REFERENCES

Blench, Roger
    2006 The Austronesians in Madagascar and on the East African Coast: Surveying the Linguistic Evidence for Domestic and Translocated Animals. Paper given at the International Conference on Austronesian Languages. Puerto Princesa, Palawan, January 17–20. Draft.

Bodin, Örjan, Maria Tengö, Anna Norman, Jakob Lundberg, and Thomas Elmqvist
    2006. The Value of Small Size: Loss of Forest Patches and Ecological Thresholds in Southern Madagascar. Ecological Applications 16(2):440–451.

Chamberlain, Ted
    2005 Photo in the News: New Lemur Species Discovered. National Geographic News, August 5. http://news.nationalgeographic.com/news/2005/08/0809_050809_lemur_photo.html, accessed November 11, 2006.

Colding, Johan, and Carl Folke
    2001 Social Taboos: "Invisible" Systems of Local Resource Management and Biological Conservation. Ecological Applications 11(2):584–600.

Cole, Jennifer
    2001 Forget Colonialism? Sacrifice and the Art of Memory in Madagascar. Berkeley: University of California Press.

Comaroff, Jean, and John Comaroff
    2001 Naturing the Nation: Aliens, Apocalypse and the Postcolonial State. Journal of Southern African Studies 27(3):627–651.

Cooke, A., J. R. E. Lutjeharms, and P. Vasseur
    2003 Marine and Coastal Ecosystems. In The Natural History of Madagascar. Steven M. Goodman and Jonathan P. Benstead, eds. Pp. 179–209. Chicago: University of Chicago Press.

Decary, Raymond
    1958 Histoire Politique et Coloniale, vol. 3: Histoire des Populations Autres que les Merina, fasc. 1: Betsileo, Betsimisaraka, Antanosy, Sihanaka, Tsimihety, Bezanozano, Antanala, Antankarana, Bara, Mahafaly, Antandroy. Antananarivo, Madagascar: Imprimerie Officielle.

Dorr, Laurence J.
    1997 Plant Collectors in Madagascar and the Comoro Islands. Washington, D.C.: Smithsonian Institution.
    2004[1966] Purity and Danger. New York: Routledge.

Environment News Service
    2006 Madagascar Declaration: Value of Nature Key to African Development. June 26. http://www.ens-newswire.com/ens/jun2006/2006-06-26-02.html, accessed November 10, 2010.

Feeley-Harnik, Gillian
    1984 The Political Economy of Death: Communication and Change in Mala-
    gasy Colonial History. American Ethnologist 11(1):1–19.

Frazer, James George
    2000[1922] The Golden Bough: A Study in Magic and Religion. III. Part
    2, Taboo and the Perils of the Soul. Elibron Classics. Facsimile reprint of
    the 1922 edition by Macmillan and Co., Ltd., London. New York: Adamant
    Media Corporation.

Gerety, Rowan Moore
    2009 Major International Banks, Shipping Companies, and Consumers Play
    Key Role in Madagascar's Logging Crisis. December 16. http://www.wild
    madagascar.org, accessed May, 5, 2010.

Graeber, David
    1995 Dancing with Corpses Reconsidered: An Interpretation of "Famadi-
    hana" (in Arivonimamo, Madagascar). American Ethnologist 22(2):258–278.

Jarosz, Lucy
    1994 Taboo and Time-Work Experience in Madagascar. Geographical Re-
    view 84(4):439–450.

Jones, Julia P. G., Mijasoa M. Andriamarovololona, and Neal Hockley
    2008 The Importance of Taboos and Social Norms to Conservation in Mada-
    gascar. Conservation Biology 22(4):976–986.

Kaufmann, Jeffrey C.
    1999 Faly aux Vazaha: Eugène Bastard, Taboo, and Mahafale Autarky in
    Southwest Madagascar, 1899. History in Africa 26:129–155.

Keller, Eva
    2009 The Danger of Misunderstanding "Culture." Madagascar Conservation
    and Development 4(2):82–85.

Kull, Christian
    1996 The Evolution of Conservation Efforts in Madagascar. International
    Environmental Affairs 8(1):50–86.

Kus, Susan, and Victor Raharijaona
    2000 House to Palace, Village to State: Scaling Up Architecture and Ideology.
    American Anthropologist 102(1):98–113.

Lambek, Michael
    1992 Taboo as Cultural Practice among Malagasy Speakers. Man, n.s.
    27(2):245–266.

L'Express de Madagascar
2002 "New York Stories." No. 2078. Thursday, January 3. P. 2.

Mittermeier, Russell A., William R. Konstant, Frank Hawkins, Edward E. Louis, Olivier Langrand, Johah Ratsimbazafy, Rodin Rasoloarison, Jörg U. Ganzhorn, Serge, Ian Tattersall, and David M. Myer
2006 Lemurs of Madagascar. 2nd ed. Illustrated by Stephen D. Nash. Washington, D.C.: Conservation International.

Mullin, Molly
1999 Mirrors and Windows: Sociocultural Studies of Human-Animal Relationships. Annual Review of Anthropology 28:201–224.

Rabearivony, Jeanneney, Eloi Fanameha, Jules Mampiandra, and Russell Thorstrom
2008 Taboos and Social Contracts: Tools for Ecosystem Management; Lessons from the Manambolomaty Lakes RAMSAR site, Western Madagascar. Madagascar Conservation and Development 3(1):7–16.

Ruud, Jørgen
1959 Taboo: A Study of Malagasy Customs and Beliefs. Oslo: Oslo University Press.

Shapiro, Warren
1991 Claude Lévi-Strauss Meets Alexander Goldenweiser: Boasian Anthropology and the Study of Totemism. American Anthropologist 93(3):599–610.

Shaw, G. A.
1896 The Aye-Aye. Antananarivo Annual and Madagascar Magazine 2.

Shoumatoff, Alex.
1988 Our Far Flung Correspondents (Madagascar). Dispatches from the Vanishing World, March 7. http://www.dispatchesfromthevanishing world.com/pastdispatches/madagascar/printermadagascar.html, accessed May 12, 2010. Also published in the New Yorker, March 7, 1988.

Sibree, James
1892 Curious Words and Customs Connected with Chieftainship and Royalty among the Malagasy. Journal of the Anthropological Institute of Great Britain and Ireland 21:215–230.

Sodikoff, Genese
2009 The Low-Wage Conservationist: Biodiversity and Perversities of Value in Madagascar. American Anthropologist 111(4):443–455.

Steiner, Franz
    1956 Taboo. Preface by E. E. Evans-Pritchard. Laura Bohannen, ed. New
    York: Philosophical Library.

Valeri, Valerio
    2000 The Forest of Taboos. Madison: University of Wisconsin Press.

Van Gennep, Arnold
    1904 Tabou et Totémisme à Madagascar: Étude Descriptive et Théoretique.
    Paris: Leroux.

Walsh, Andrew
    2002 Responsibility, Taboos and the "Freedom to Do Otherwise" in Anka-
    rana, Northern Madagascar. Journal of the Royal Anthropological Institute
    8(3):451–468.
    2005 The Obvious Aspects of Ecological Underprivilege in Ankarana,
    Northern Madagascar. American Anthropologist 107(4):654–665.
    2006 "Nobody Has a Money Taboo": Situating Ethics in a Northern Mala-
    gasy Sapphire Mining Town. Anthropology Today 22(4):4–8.

Wildmadagascar.org
    2005 New Lemur Species Named after British Comedian. November 12.
    http://news.mongabay.com/2005/1112-lemur.html, accessed September 6,
    2006.

Part 2. Endangered Species and Emergent Identities

4. TORTOISE SOUP FOR THE SOUL

FINDING A SPACE FOR HUMAN HISTORY IN EVOLUTION'S LABORATORY

Jill Constantino

Categories of Belonging

When we swat flies, eat dolphin-safe tuna, use bug spray, or give money to protect pandas we are deciding which nonhuman beings belong in particular places and which do not. When we fill our universities, issue travel visas, consider the land rights of indigenous people, or prohibit the passage of immigration laws, we are making decisions about human belonging. What are the factors that influence "belonging"? How long must a being exist in one place in order to belong? Do creatures belong after a quantifiable period of time, or is their belonging more dependent on qualitative factors like being the first to a place, being among the last in a place, being unique to a place, or claiming an origin myth involving that place? What characteristics must a being exhibit in order to be protected or eradicated? Clearly, belonging is subject to various cultural factors and scientific findings.

The variables and characteristics that form categories of value differ from species to species and emerge from different time frames. For nonhuman beings, evolutionary time provides a context through which to determine endemism or native status. For humans, historical time provides a ground for the construction of indigenous identities, often connoting special value and special rights. But what happens when the contexts of human and nonhuman creatures merge? In nature reserves, national parks, coastlands, farms, logging sites, and even in our cities, how might we decide which beings have the right to the resources and the privileges of the places they inhabit? When people craft their own identities of value in the arbitrary constructions of belonging, how do they negotiate between and among the frames of evolutionary time and historical time?

In this chapter, I offer examples of these negotiations in the Galápagos Islands, which are celebrated for their evolutionary processes, famous for their iconic species, and prominent in the history of science. In the Galápagos, where conservationists work to preserve the lives of species made famous by Charles Darwin, human residents engage in a politics of staking historical and geographical claims to legitimate social belonging. They do this in relation to highly valued nonhuman creatures. The tensions between conservationists, who advocate for the well-being of endemic and endangered creatures, and Galápagos residents, who struggle for recognition in the archipelago, draw on frames of reference that blur the boundaries of evolutionary and historical time. Since human populations in the Galápagos date back less than two hundred years, residents do not identify as indigenous, nor do they campaign for the rights associated with such an identity. Instead, some have strategically aligned themselves with high-status species around which conservationists rally, such as Lonesome George, the last remaining tortoise of his subspecies. Others work to establish a folk category of endemism as a means of legitimating their claims to place. Since nonhuman creatures can evolve in shallow time, they suggest, why can't multigenerational human residence in one place achieve endemism? Through the Galápagos case, I highlight the value of endemism in conservation discourse, the arbitrariness inherent in value systems of belonging, and the social construction of taxonomic groups.

This chapter emerges from my experiences and research starting in 1995–96 and including return trips to Ecuador and the Galápagos in 1997, 1998, 1999, 2000, 2001–2002, and 2006. I first came to the Galápagos in 1995 as an English teacher, eager to see the famous flightless birds, swimming reptiles, and giant turtles of Galápagos and largely unaware that any humans lived in the Galápagos at all. I did some preliminary research on the community I would live in, finding only one brief mention of the human populations on Isabela Island in a guidebook. Once on the islands, I met a scientist at the Charles Darwin Foundation Library who responded to my inquiry about the lack of writing on humans with "Don't you know, we're trying to keep people out!" As I made friends and found fictive family in the Galápagos, I became more intrigued by the absence of humans from Galápagos writing and by the social strife fomenting at that time. In 2001, I returned to the archipelago as a Fulbright Scholar, eager to explore the social conflicts and to introduce humans into Galápagos writing through my own anthropological work. In this chapter, I write about my research findings using the present tense because time is of the essence for human and nonhuman populations in the Galápagos;

the present tense gives a sense of immediacy to inspire thoughtful attention to the problems of all Galápagos creatures.

Galápagos Groups

Scientists, conservationists, and tourists follow Charles Darwin's footsteps through the Galápagos Islands; they walk the beaches, climb the lava rock, and dive through the ocean in hopes of finding what Darwin glimpsed in the archipelago, that "mystery of mysteries," or the particular matrix of space and time in which species originate (Darwin 1964:1). People interested in Galápagos nature consider the arrival of new beings and the departure of beings, the spread of creatures and the disappearance of creatures, how living things change and how these changes affect the complex web of interconnected organisms. Why and how did organisms vary, adapt, and find their niches in their new worlds? The tortoises, iguanas, cormorants, and finches are internationally celebrated as iconic subjects of evolutionary process. Their unique behavior, contributions to Galápagos biodiversity, and role in the development of Charles Darwin's theory of natural selection contribute to their fame. The 25,000 humans who live in the Galápagos are less celebrated.

Humans didn't arrive in the Galápagos until the fifteenth century. Explorers, conquistadors, buccaneers, and whalers came to the volcanic rock as they guided their vessels through the strong currents of the archipelago. Settlement colonies formed in the nineteenth century and expanded through the twentieth with tourism opportunities and new fishing markets. Humans brought ashore new forms of life in the treads of their shoes, the waste of their bodies, and the cargo of their ships. These introduced organisms entered the Galápagos ecosystem and the competition for resources on the archipelago.

Today, scientists and conservationists group island populations into species that belong on the archipelago (those endemic and native species that arrived and evolved without human intervention) and species that don't belong (those introduced and invasive organisms that arrived with humans). People from the Galápagos National Park and the Charles Darwin Research Station conserve native and endemic organisms in their "living laboratory" of evolution through carefully considered and studied processes (Charles Darwin Foundation n.d.). They set up breeding colonies, protect food sources, and actively control nonnative and nonendemic populations that create competition for the islands' resources. They cull invasive plants that have entered with the materials of human infrastructure. They exterminate rats, cats, and dogs that have arrived through human transportation systems. And they shoot goats that threaten endemic populations. The practice of controlling these interdependent relationships is a direct expression of power not lost on the people who live in the Galápagos.

Fishermen and their families from Isabela Island wonder if their futures on the archipelago are certain, if they might be the next species to be controlled, culled, or eradicated. People there, called *Isabeleños,* claim more than seven generations of history. Their narratives ignore, dispute, and incorporate the intellectual and scientific processes that have made the Galápagos animals iconic on an international stage. Conflicts mount as Isabeleños, under pressure to conform to conservationist ideals, nonetheless fish among depleted stocks both within and outside of sanctioned waters and seasons. All groups of people on the island struggle to find appropriate places for themselves and for other organisms in the shifting contexts of Galápagos evolution and history. Even amidst conflicts, most residents, visitors, and admirers of Galápagos life embrace the image of the Galápagos as an example of pristine nature, relatively free of human incursion. They thereby accept the categories of science and conservation and divide organisms into those that belong and those that don't belong.

How, then, do humans who live in the Galápagos find legitimacy when they do not belong? How do they connect their histories to the categories of value—to the groups of organisms that belong? The guarded and endangered giant tortoises, being the largest and arguably the rarest beings of value, figure centrally. Fishermen position themselves as endemic or native to gain the legitimacy that tortoises possess or, at the least, to criticize and restructure the categories of value that ideologically dehumanize and logistically restrain human lives. Galápagueños align themselves with the good creatures (the tortoises, the iguanas, and the finches) and against the bad (the goats, rats, cats, and frogs); they control tortoise habitat and they monitor and intervene in tortoise reproduction. Professional and amateur scientists demonstrate their knowledge of biological systems and beings and thereby offer their services to the maintenance of scientific categories.

While we may be tempted to think of the archipelago as a humanless timescape of natural evolution and even yearn for the creation of such a place, such understandings of human and nonhuman systems as separate are misleading and dangerous. These understandings naturalize categories of value and thereby create implications of human worth and worthlessness. The implications in turn feed conflict, impede the processes of conservation, and violate human rights. When people choose to reject the categories with which others align themselves by advocating for human rights in the archipelago, conflicts often culminate in violence to both the animals and the humans of the Galápagos.

Identifying with Tortoises, Cows, and Goats

The Galápagos offer a particularly salient model for study because the processes of conservation and evolution are coffee-table talk in the archipelago and conflicts over natural resources are central to everyday life. Residents often identify with their nonhuman kin in tribute to the iconic place of evolutionary processes: local children in theater groups align themselves with nonhuman creatures as they become sea lions, frigate birds, and iguanas in performances for international tourists; friends take on animal nicknames like Iguanaman and *Lobo* (sea lion); and many children are named Darwin—the man who "netted" it all together. People identify with the Galápagos creatures or the processes of evolution through playful names and simple symbolism, but they also identify with their nonhuman kin as organisms worthy of the islands' resources.

In 2001, fishermen from Isabela Island positioned themselves alongside Galápagos tortoises as creatures deserving of medical care. When a rumor circulated that the Charles Darwin Research Station was considering closing the health clinic on Isabela to encourage emigration, fishermen and their families who depended on that clinic were angry but not surprised. They recalled other equally infuriating incidents. In the mid-1990s, the research station reportedly airlifted a Galápagos tortoise to Miami for a paw operation. In that same year, three Galápagos fishermen died in diving-related accidents who might have lived if conservationists and scientists had collaborated in obtaining essential medical resources and safety equipment. People from Isabela wondered how those from the national park and the research station could justify such extravagant expenditures on nonhuman species while neglecting members of their own species. While Isabeleños reject the science that they see as valuing the nonhuman over the human, they strive to legitimate their lives and histories in the eyes of those scientists and conservationists who, as many on the islands see it, control the resources and policies of the archipelago.

In the process of legitimization, a group of Isabela men sought to identify themselves as endemic through an expedition to find a small cow. Wild cattle breed among the scrubby vegetation which grows from the volcanic soil on the southern slope of Cerro Azul, one of the most active volcanoes in the Galápagos archipelago. A particularly small type of longhorn cow, not much larger than a German shepherd, is rumored to exist amidst the aridity, having adapted to the extreme conditions. In February 2002, the Isabela residents set out to find the *vaca-perro* (cow-dog), and, by doing so, to prove that they themselves were creatures of value, like Galápagos tortoises. Conservationists generally consider cows invasive and thus deserving of eradication. The fishermen wanted to find evidence that a species of cow had in fact evolved on the island to the point where it should be considered endemic and worthy of conservation. If it is true that species evolve quickly in circumscribed island environments, the Isabela men reasoned, then perhaps the bad organisms, often categorized as nonnative, nonendemic, and invasive, might be considered good.

The group of Isabeleños proposed that the discovery of a miniature longhorn could provide a case for human endemism within a historical time frame and, therefore, valorize these human lives and histories in a conservationist context. In the end, the group reported finding the evidence they were looking for—a very small longhorn skull. They also reported that, unfortunately, their key evidence had disappeared shortly after they found it. A torrential downpour swamped their fiberglass boat on the evening of the discovery and washed its contents to the bottom of the ocean. Despite this, by adopting the scientific category of endemism and by manipulating the context in which this category is used, people from Isabela challenged the principles on which the Galápagos National Park and the Charles Darwin Research Station base their conservation efforts.

While fishermen and their families challenge, manipulate, and reconstitute the categories of science and conservation, so do scientists and conservationists. Nonhuman species are vaulted to superhuman status; tortoises are anthropomorphized into dynastic rulers. In a popular book published by the Charles Darwin Foundation in 1999, British biologist and longtime Galápagos resident Godfrey Merlen writes about the intertwined human and tortoise histories and the importance of tortoises to Galápagos conservation. In this passage, we see the tortoises' iconic status as a piece of threatened Galápagos nature.

> Even before the Research Station's inauguration, the Research Station's first conservation officer, Miguel Castro, was out in the field searching for the beleaguered tortoises, trudging over the lava fields, bursting through the spiny vegetation, thirsty, hot, and tired. It was, in a sense, a search for the Holy Grail, for the tortoise is the very symbol of Galapagos and its conservation. Without them, what was there to hang on to? (Merlen 1999:11)

Merlen's book on terrestrial turtles, their close call with extinction, and their quasi-religious historical importance is entitled *Restoring the Tortoise Dynasty* (1999). Given this title, we might imagine a succession of reptilian rulers who, having long dominated the Galápagos, were ousted from power by humans—a power that the tortoises rightfully held and that must be "restored." Humans are both the ignorant tyrants of this tale and, then, the repentant heroes who right their wrongs and set the natural world back on its proper course.

The Isabela Project, an internationally funded, multimillion-dollar eradication effort, offers an example of "restoring" order. Since 1997, people from
the Charles Darwin Research Station and the Galápagos National Park have
eliminated goats from the northern half of Isabela Island. The introduced
goats ate the food of endemic tortoises, decimated native plant populations,
and caused erosion on the pristine volcanic slopes. To find and kill the goats,
park and station workers used helicopters, .22 caliber rifles with telescopic
sights, geographic information systems, trained dogs to find and herd the
goats, trainers to train these dogs, facilities to house them, tiny dog boots to
protect paws against jagged volcanic rock, and "Judas goats"—goats with radio
collars who betray their goat kin by leading hunters to herds.

Knowing the rough details and economic scope of the project, people who
live on Isabela Island were bothered by the international money that went into
hunting these goats; they wondered why some of that money couldn't come
to the local populations, whose health care and school systems were failing.
And they wondered why they couldn't, at the least, eat those goats that lay
rotting on the volcanic slope on the other side of the island. The great concern
for the endangered Galápagos tortoises and the lack of perceived concern for
the problems of human populations once again illuminate the conflicts of sustainable living that happen among species with differently conceived rights of
belonging. The image of people firing machine guns from helicopters at elabo-
rately herded goats illustrates conservation around categories of belonging, the
violence of conservation projects, and the fear that this violence might bring
to those who live nearby and who find themselves outside of the categories of
belonging.

While Isabeleños won't find themselves in the scopes of machine guns
pointing down from helicopters above, the policies and processes of science
and conservation often constrain their daily practices and reconstruct their
histories. Scientists and conservationists erase and manipulate Isabela pasts as
they craft images of a pristine archipelago or a threatened one—images that
draw in research dollars and conservation support. Isabeleños consider their
futures far from certain. Regulations and sanctions constrain their livelihoods,
requiring them to fish and harvest less often for fewer fish that bring in smaller
incomes.

On the Wrong Side of Natural History

When people from Isabela refuse to align themselves with the tortoises
or other popular Galápagos creatures, they find themselves on the wrong side
of natural history, and as a result they are demonized in the international me-

dia. In September 2001 in the middle of the night, fishermen from Isabela Island (the center of fishing efforts in the Galápagos) gathered up several giant tortoises from the corrals in the Charles Darwin Research Center on Isabela. They moved the tortoises (with great effort—they are very heavy) to the National Park Offices a few blocks up the sand road, where they built a crude enclosure and then planted themselves alongside the tortoises in an act of protest. The fishermen hoped to bring attention to their human plight. As they see it, research station scientists and national park employees give more rights to tortoises than they do to Galápagos people. During my fieldwork at the time of these protests, Isabela fishermen explained that they never intended to cause any harm to the tortoises. They just wanted to draw attention to their struggles. Perhaps, if scientists, conservationists, and outsiders paid greater attention to their lives and their histories, they might understand and appreciate the human difficulties of living on such a celebrated archipelago in the Pacific.

On the first day of the protest, one fisherman—who told me he was trying to be funny—posed with a knife at the throat of a tortoise. This pose and its implicit threat were widely reported in newspapers, magazines, and radio programs. The protest struck an international nerve. When fishermen used Galápagos tortoises in their protest, they hoped to broadcast the absurdity of valuing human/nonhuman relationships over human/human relationships. They implicitly and publicly asked, "Are turtles really more important than fishermen?" The international answer was a resounding "Yes!"

Other than by actions that might concern conservationists and provoke calls for greater regulation of human populations, people on the islands have a difficult time claiming a place for themselves at all. Human populations on the Galápagos only date back 150 years. The absence of predator populations in deeper time created the striking evolutionary scene that Darwin wrote about. People, as predators, pose a direct threat to this evolutionary scene. Further, the nonhuman creatures that inhabit that scene, set in deeper evolutionary time, have already been popularly celebrated as the creatures that belong. Humans must compete with the tortoises, iguanas, and finches that have already claimed their place in Galápagos history through the theories of Charles Darwin, conservation literature, and ecotourism. These public histories demand that humans play only a certain role, or no role at all. While Isabela fishermen attempt to make a case for their rights relative to the rights of the more famous beings that populate the lava, the giant tortoise narrative is particularly tricky to enter.

Fishermen enter this established Galápagos history through two avenues; they enter through the linear structures of reptilian power and ancestry set up in books like *Restoring the Tortoise Dynasty* or they enter as organisms in ecological cycles. Entering as part of the ecological cycle is difficult because the form of such a cycle doesn't necessarily fit the unidirectional drop in tortoise numbers that accompanied early human history in the Galápagos, or the unidirectional increase in tortoise numbers resulting from contemporary conservation work. We might consider the drop and increase in tortoise numbers to be a fluctuation of a cycle triggered by the introduction of new species, but then fishermen are easily singled out as an invasive species, predators, and opportunistic outsiders—negative labels that further dehumanize. And this positionality highlights the devastation of tortoises at the hands of humans. Although buccaneers, whalers, and scientists of the past might be held accountable for the near extinction of tortoises, an ecological cycles perspective positions all humans, including present-day fishermen, as responsible.

When fishermen force themselves into the more historical dynastic trajectories of tortoises through protest actions, their efforts backfire. When Isabela fishermen took tortoises hostage, for example, they were trying to make a case for their human needs, but instead they represented themselves as insurgents bent on toppling a dynasty. Positioning tortoises as powerful and rightful rulers of the archipelago and humans as momentary insurgents helps many conservationists promote their idealistic goal of keeping the islands pristine—free of human incursion—and facilitates fund-raising for the purpose.

Loving Lonesome George

Scientists, conservationists, and tourists more easily enter the tortoise narrative. Wooden bridges and walkways connect tourists to different tortoise corrals. Visitors walk among the reptiles, feel their enormity, and pose for photos next to the slow pieces of the past—a past imagined as located before the decimation of the tortoises, before development, when "nature" was still pristine, or even as prehistoric (these tortoises have a dinosaur look to them). The huge animals in the corrals might be a hundred years old—maybe even older. It is possible, tourists openly speculate, that Charles Darwin walked on the islands alongside these very creatures. The tourists witness nonhuman nature, depart from the islands, and leave their money behind—often specifically earmarked for Lonesome George.

While the Tortoise Breeding Center on Santa Cruz has a largely hopeful feeling, with new tiny turtles scurrying around in their protected pens and the older ones marking their escape from extinction, the "sad state" of the "magnificent creatures" that Merlen describes in *Restoring the Tortoise Dynasty* is most powerfully embodied—or, a better word, *personified*—in Lonesome George, "the rarest living creature," according to the Guinness Book of World Records. Lonesome George's corral has been marked with placards that explain his fate and the bleak possibilities for his future. He may be the last of his subspecies of giant Galápagos tortoises. Since a national park worker found him in 1971 (a full 60 years after any other tortoise sighting on Pinta Island), the search to find George a partner of his own subspecies has been unsuccessful. Biologists test his semen and his diet, assessing his reproductive potential. They attempt to pair George with other morphologically similar species from other islands to preserve Pinta DNA in the Galápagos tortoise gene pool. In *Restoring the Tortoise Dynasty,* Merlen writes, "George is, of course, a concern to everybody. He is the unquestionable symbol of all that can go wrong" (1999:50). When George dies, his line of inheritance will end and his subspecies will be extinct.

Researchers have considered cloning Lonesome George. The process would be complicated and expensive. Merlen explains the process in *Restoring the Tortoise Dynasty.* With no eggs or embryos to work with, scientists would need to implant genetic material into an empty egg—one with its own genetic information removed—place it within a surrogate mother until the eggs mature and are laid, and then maintain the eggs at a certain temperature to produce a male or female (Merlen 1999:51). George's photo, alongside a description of the cloning process, its difficulties, and its potential results, decorates the breeding center grounds and research station literature. Lonesome George is the anthropomorphized poster boy of Galápagos conservation. If scientists successfully clone George, they will be responsible for the continuation of this particular line of the dynasty. They will plant themselves firmly in his historical line and establish an acceptable place for themselves in the Galápagos.

Scientists who work to continue his subspecies directly enter his reproductive world either through their scientific processes or through their talk. When biologists, conservationists, and tourists discuss Lonesome George and his sad state of affairs, his sex life is a frequent conversation topic; they connect to George by thinking about him having sex, speculating on worthy partners, and even stimulating him to ejaculation. Research station and park employees placed mating turtles in front of Lonesome George in hopes of inspiring him to follow their example. They offered reward money to anyone who might find him a suitable partner. They directly enter his dynastic line. A national park worker explained to me that George had been "in love" with a herpetologist. He explained that the herpetologist approached George in her best turtle posture and eventually used her hand to sexually stimulate him. People talk about how they "set him up" with two "fine" females from another island, how George "hooked up" but didn't make any babies. A particularly evocative Lonesome George sex narrative comes from the Tortoise Trust, an internet group devoted to caring for and protecting tortoises:

> I came to be leaning on the gate of George's corral to witness the first signs of passion . . . The female from Volcan Wolf was having a mud bath when George lumbered into action. "This is it," I thought . . . He moved purposefully behind her and nudged her shell with the front of his own . . . Startled, she climbed quickly out of the wallow. "Go for it, George," I whispered. He plunged into the mud-hole . . . and stayed there. "Oh, George!" [Seal n.d.]

In the *New York Times*, John Tierney writes about his "usual impulse to fix up the world's most famous bachelor":

> As I fought my way through Pinta's overgrown vines and cactus pads, fervently hoping to spot a female Pinta tortoise behind a lava rock or a thorn bush, I was already working on her name. Georgette seemed too derivative. I liked the local evolutionary allusion of Darwinia, but finally settled on less of a mouthful: Eve. (Tierney 2007)

If people can help George mate through their direct stimulation, romantic nudging, or matchmaking adventures, they might be responsible for saving George's subspecies. Their presence in the Galápagos narrative becomes legitimate as they attach themselves positively to the tortoise dynasty and to the reproductive line that runs back through evolutionary time.

Fishermen and their families also consider the reproductive potential and abilities of Galápagos organisms and thereby enter the dynasty. A longtime Isabela resident told me about tortoise reproduction. He asked if I knew that the New York City Garden had a Galápagos tortoise. Although I can find no record of a New York City Garden that presently holds a Galápagos tortoise, he nevertheless explained that the tortoise refuses to reproduce there; it isn't the right climate. They only reproduce in the Galápagos, he said. Perhaps he was using his position as a Galápagos inhabitant to make a case for his connection to the dynasty. He recited many facts about Galápagos reproduction that day. He explained that invasive frogs, a target of station and park eradication efforts, produce far too many eggs and are too tiny to be effectively picked from the water. Displaying his folk knowledge of Galápagos fauna, he positioned himself as an actor in population trajectories. He may have been carving a place for himself and for his history by entering those dynasties—dynasties with narratives that someone like me, a person from the United States with a science background, might hold central.

Tortoise Soup

Generally, though, fishermen and their families on Isabela Island refuse to speak and act for Galápagos tortoises or other famous Galápagos creatures. Instead, they advocate for themselves. They worry about their dwindling opportunities, lack of resources, and lack of food. On a clear hot day in 2001, I talked with Pablito and Efrén, the elderly sons of Captain Serapio Jaramillo, who was exiled to Isabela Island in 1918 as a political prisoner. They were sitting in two straight-backed chairs on the cement stoop of Pablito's house. Pablito's wife, Esperanza, was inside in the hammock, interjecting important details when necessary. I sat with my feet in the sand street and we all looked out over the little brackish pond on the periphery of Puerto Villamil. As they spoke, a handful of flamingos were picking through the pink mud. Although, as the brothers told me, hundreds of giant Galápagos tortoises used to cool themselves in that very pond, they were long gone by the time Pablito and Efrén got there. Pablito and his brother talked about the constantly increasing population, accompanied by the diminished returns from the sea and the restrictions on agriculture and fishing. "What will the people have left to eat?" Efrén asked rhetorically. With a grin, Pablito offered, "Los Galápagos!"—the Galápagos tortoises. At that moment, a half-dozen tourists walked by—a rare but increasingly common occurrence in Puerto Villamil. At the sight of the tourists in front of his brother's house, Pablito exclaimed, "Turista a la plancha!" (grilled tourist). Efrén delightedly added, "Turista al vapor!" (steamed tourist).

Those attached to the narrative that depicts the Galápagos as the last pristine place would like fishermen and their families from Isabela to be a bit more silent. In order for conservation and tourism to work in the Galápagos, humans must have only minimal impact, while nonhumans thrive. The Galápagos are famous for an evolutionary process distinctly unmarked by high-order predators, including humans. Tourism works as tourists feel themselves walking in Darwin's footsteps and seeing what the world might have looked like before humans. People infringe on this image and, indeed, change Galápagos ecosystems through their fishing, their hotels, their infrastructure, and the species they introduce. Conservationists and tourism workers attempt to minimize their own presence. They represent themselves as existing on the islands only in order to conserve them through their research, their programming, and their ecotourism. When fishermen vocally and publicly advocate for their own rights without appropriately aligning with creatures that belong or without finding a niche for themselves in the categories and vocabulary of science, they push against the idea that the Galápagos human presence should be minimal or even silent.

When fishermen advocate for their own rights by using and even threatening harm to the beings around which others silence themselves, they commit a double offense in the hegemonic Galápagos moral order. In this moral order, the problem is not only that fishermen are overfishing the marine reserve, or that they fish illegally, but that they don't seem to accept the Galápagos founding legend of the Tortoise Dynasty, set in historical time, nor do they accept the scientific frames of evolutionary time which delimit categories of belonging. They don't seek to belong by aligning themselves with the iconic creatures of the Galápagos.

Negotiations through participatory management systems may be helpful in mediating and ameliorating seasonal surges in conflict in national parks and nature reserves. But if humans are only supposed to speak for and align with nonhuman others, and aren't allowed to speak for themselves as people positioned in both historical and evolutionary time, it becomes impossible for humans to speak appropriately for one another. Further, without considering the arbitrariness of categories of belonging and the blurred boundaries of the time frames which uphold them, we will not be able to fully disentangle the value systems at work in places like the Galápagos, where humans and nonhumans meet. Until human history is publicly acknowledged and until all people have a right to their history, no matter how short or long, negotiations over natural resources will result in frustration. Fishermen, in ongoing protest, will continue to make tortoise soup as they did when they first arrived on the islands, before they were silenced by the more powerful narratives of science and conservation.

Epilogue

Lonesome George's state of affairs recently took a bright turn. In an article sent out through Reuters, we learn that "Lonesome George may end [his] bachelor days on the Galápagos." George "stunned his keepers" by mating with one of the closely related tortoises in his pen; several tortoise eggs appeared shortly thereafter (Soto 2008). While biologists in the turtle nursery eventually released the disappointing news that the eggs were infertile, George's willingness to mate is a positive sign. Even better news comes from Isabela, where DNA tests reveal Pinta blood in the small tortoise gene pool. These results give hope that Pinta individuals exist on Isabela (Russello et al. 2007). Lonesome George may not be alone. Others of his subspecies may be slowly pulling themselves along the rim of Wolf Volcano. The giant Galápagos tortoises might be enjoying the rebounding vegetation that grows amidst the bones of goats, gunned down from helicopters above.

REFERENCES

Charles Darwin Foundation
N.d. "Who We Are: About Us." http://www.darwinfoundation.org/english/
pages/interna.php?txtCodiInfo=73, accessed August 31, 2010.

Darwin, Charles
1964[1859] On the Origin of Species: A Facsimile of the First Edition. Cambridge, MA: Harvard University Press.

Merlen, Godfrey
1999 Restoring the Tortoise Dynasty: The Decline and Recovery of the Gala-
pagos Giant Tortoise. Quito, Ecuador: Charles Darwin Foundation.

Russello, Michael A., Luciano B. Beheregaray, James P. Gibbs, Thomas Fritts,
Nathan Havill, Jeffrey R. Powell, and Adalgisa Caccone
2007 Lonesome George Is Not Alone among Galápagos Tortoises. Current
Biology 17(9):R317.

Seal, Vicki
N.d. Lonesome George. http://www.tortoisetrust.org/articles/george.html,
accessed October 14, 2010.

Soto, Alonso
2008 Lonesome George May End Bachelor Days on Galápagos. Reuters, July 22.

Tierney, John
2007 A Lonesome Tortoise, and a Search for a Mate. New York Times, May 8.

5. GLOBAL ENVIRONMENTALISM AND THE EMERGENCE OF INDIGENEITY

THE POLITICS OF CULTURAL AND BIOLOGICAL DIVERSITY IN CHINA

Michael Hathaway

During the 1990s, some scholars and activists invented a new position, bridging what used to be fairly separate realms of indigenous rights and environmentalism. They argued that biological diversity and cultural diversity were vitally important, threatened, and connected. Although the link between the protection of the environment and of the rights of indigenous peoples might now seem obvious, alliances between environmental and indigenous rights groups were new at that time, first gaining momentum during the contentious politics of the Amazon rain forest in the 1990s (Conklin and Graham 1995). In this frame, particular peoples and particular natures were both said to be threatened with extinction. In turn, such alliances have shifted the terrain for environmental and indigenous rights efforts around the world, fostering new activist groups, pushing environmentalists to foreground better relations to indigenous peoples, and prompting indigenous rights organizations to argue on environmentalist grounds.

This chapter explores one outcome of such connections, whereby international concern about habitat and species loss in China has played a part in the emergence of a new indigenous identity. The issue of indigenous identity in China is remarkable and complex because the Chinese state declares that the concept of indigenous people is irrelevant there. To officials in Beijing, all the nation's peoples are equally indigenous. When I was in China in the mid-1990s, there was little idea of valorizing the "primitive," either as a "noble savage," a positive alternative to corrupt society, or as an "environmentally noble savage," an example of an ecologically sustainable lifestyle (Trouillot 1991). Almost all groups that could potentially be seen as indigenous by international frameworks, such as those who lived far from major urban centers and carried on activities regarded as "traditional" by a Western framework, were often described in China as backward (*luohuo*), lacking in culture (*meiyou wenhua*), and dirty (*zang le*) (Schein 2000; Litzinger 2000; Mueggler 2002). Yet this was also the period in which Chinese scholars and activists were beginning to open up what I call an "indigenous space," mostly through their engagement with environmental projects. These scholar-activists[1] have not tried to import the concept of indigeneity writ large, but instead engage in the more contingent and careful work of recrafting the concept within powerful and resistant political spaces.

This chapter comprises two sections. First, I briefly sketch a history of how notions of cultural and biological diversity became linked and tied to environmentalism and indigenous politics. Second, I discuss how indigeneity has emerged in China in reference to these environmental contexts, looking at how two Chinese scholar-activists attempt to make the indigenous space a political reality.

Linking Cultural Diversity and Biological Diversity

Before the 1990s, most proponents of protecting biodiversity (a term itself only invented in the 1980s) regarded themselves as advocates for plants and animals. Such proponents often encouraged governments in developing countries to create and patrol nature reserves, even if this entailed the relocation of indigenous people (Conklin and Graham 1995). On the other hand, advocates for cultural diversity argued that indigenous peoples were threatened by government policies, corporate acquisition of land and resources, and even environmental efforts to create reserves.

In the 1980s, indigenous rights groups began to borrow strategies from environmentalists. In particular, they borrowed new frames such as the discourse of extinction and endangerment. Increasing numbers of films, books, and projects rallied around the concept of "endangered peoples" or portrayed such groups as parts of "disappearing worlds." Although, at first glance, this move seemed to repeat concerns raised a century earlier about "vanishing Indians," there are some critical differences (Brosius 1997). Such a "vanishing" was earlier regarded as inevitable, but the contemporary use of the term "endangered" is intended to provoke action. Now, the word "endangered" is not fatalistic but instead suggests the promise of cultural preservation or renaissance. In the last few decades, indigenous groups have connected with each other and gained global visibility and power (Kearney 1995). Indigenous groups and their allies now form a growing presence and have gained new sets of legal rights recognized by global institutions such as the United Nations and the World Bank (Niezen 2003).

The Amazon rain forest of the 1990s was one of the key places where members of the indigenous rights movement and the environmental movement came together (Turner 1991; Sponsel 1995). The Amazon captured the world's attention as a site where cultural and biological diversity were being destroyed by massive logging, agricultural, and dam projects, as well as by the influx of millions of urban migrants searching for gold and land. Over the decade, members of environmental and indigenous rights groups brokered a number of alliances around a number of common causes. Through new studies and maps, they were able to offer evidence for the first time that places of high biological diversity often overlapped with places of high cultural diversity. These findings bolstered the claim that environmental protection could be achieved through the political support of indigenous people's right to manage land and resources (Brosius 1997).

Sensibilities did not shift only within the academy. More importantly, indigenous groups put pressure on international environmental organizations to stop their long-standing practices of displacing "local people" and obstructing their access to land and natural resources (Dowie 2009). In turn, international environmental projects are now expected and required to show interest in indigenous knowledge and rights in both their conception and their execution.

Yet when foreign environmentalists came to China in the 1990s, they encountered a country in which the concept of "the indigenous" had little resonance. Almost no one in China identified as indigenous themselves, and no one championed indigenous rights. The concept appeared, as the state insisted, completely irrelevant to China. The Chinese state's position is not unique; a number of other governments, including those of India, Indonesia, and Russia, share the view that the concept of indigeneity is not applicable to their citizenries (Gray et al. 1995). Nevertheless, the indigenous category has recently gained momentum in each of these nations, albeit in very different ways (Tsing 1999; Li 2000; Gray 2005). China's past has shaped notions of cultural change and endangerment in specific ways, and these have informed the ways in which indigeneity is politically deployed and understood.

Culture, Eradication, and Extinction

During the 2008 Olympics in Beijing, there was much celebration of the ancient and rich heritage of "traditional Chinese culture." Yet for many older

Chinese, public portrayals of the glorious past are relatively novel. Under Mao Zedong's leadership, from 1949 to 1979, the idea that a rich cultural heritage had to be saved from extinction did not find much traction. The concept of "eradication," in contrast, has been far more powerful over China's past century. Unlike "extinction," which often evokes a sense of regret and implies a lack of intentionality, the closely related term "eradication" often implies a sense of achievement and purpose. One of China's more recent and far-reaching attempts at cultural eradication occurred during the Cultural Revolution between 1966 and 1976. During this time, numerous campaigns, directly sponsored by state officials or even impromptu, attempted to "destroy the four olds": old customs, old culture, old habits, and old ideas. Throughout the country, young adults, especially those who joined the Red Guards, sought out examples of the four olds, now regarded as "feudal" or "superstitious." Throughout the decade, thousands of people and objects, including statues, books, and buildings, were seen as elements of Chinese culture that should be eradicated in order to move China forward.[2] In contemporary China, such eradication campaigns are not celebrated, but often regarded as deeply regrettable. Now, people are more prone to talk about the fear of losing cultural diversity and the need to preserve certain aspects of culture.

In English, the term "extinction" has expanded its domain over the centuries; initially referring only to animal species, it now includes the realm of human culture. In Chinese, however, there are two distinct terms. *Xiaomie* (消灭), literally "to extinguish or turn off," refers to the extinction of living creatures, whereas *xiaowang* (消亡), literally "the death of a person," describes the extinction of culture, art, and other human-created forms. Recent discourse about xiaowang, or cultural extinction, which only gained traction in China in the late 1980s, is focused on four main topics: debates about World Heritage sites,[3] folk culture,[4] language extinction,[5] and indigeneity. Questions of xiaowang and indigeneity in China represent new ways in which particular groups are wrestling with notions of cultural loss, group rights, and the politics of difference.

The term "indigeneity" here refers to the politics of indigenous identity, in particular the attempt to link specific groups with a powerful global movement for indigenous rights. Therefore, I do not see "indigenous people" as a natural category defined by a timeless essence. Scholars who see indigenous people as a natural category often assume that indigenous status is clear-cut and obvious, and that indigenous struggles are mainly to get the state to recognize their status (Miller 2003). Rather, I approach the term "indigenous people" as a politicized social category. Some groups, such as the Maasai or the Kalahari San, have only recently articulated themselves as part of a global indigenous network, a linkage that offers them access to moral, political, and financial assistance (Sylvain 2002; Hodgson 2002). I trace how the social category "indigenous people" comes into existence in particular times and places, especially in seemingly ambivalent contexts, such as contemporary China.

Until recently, not only did the Chinese government assert the irrelevancy of indigenous status in China, but also ethnic groups did not identify themselves as indigenous. Global assumptions about what indigenous status entails, such as traits like tribalism and cultural isolation, have led many Chinese and foreign scholars to debate whether or not the concept of "indigenous people" is relevant to China. Whereas in the 1980s indigeneity was completely absent from the political landscape of China, by the 1990s the concept had started to emerge.

The fact that indigeneity in China has been slow to take root stems in part from China's relative isolation from the global indigenous movement. In Latin America, advocacy organizations such as Cultural Survival and Survival International played an important role in indigenous rights advocacy and in fostering regional and global connections between indigenous groups, yet such organizations have been officially unwelcome in China. Unlike citizens of the neighboring countries of Thailand, Taiwan, and Japan, who have attended international indigenous forums for decades, Chinese ethnic minorities have been notably absent from them.

This situation presents a couple of ironies. First, the Chinese state identifies itself as a supporter of indigenous rights, yet it only supports indigenous rights *elsewhere*, such as in the Americas, not in its own territory. China joins with its neighbors in saying that the concept does not pertain to Asia. Many Asian and African governments proclaim that the concept of indigeneity only makes sense for regions that were conditioned by European salt-water colonialism and permanent settlement in the colonies. From this perspective, indigeneity is applicable to Latin America, North America, Australia, and New Zealand, but not to Asia or Africa.

Second, in response to these claims by Asian and African governments, the United Nations argued that groups should be identified as indigenous on the basis of their own "self-definition" (Muehlebach 2001). The UN assumes that such self-defined status will be self-evident, yet my fieldwork revealed to me that this is not always the case. I am not aware of any groups today in China that are advocating for their own inclusion within the global indigenous network, and overall, few people attempt to employ the power of the term to describe ethnic minorities. Rather, only a small cohort of what I call scholar-activists are arguing for the relevance of the indigenous category in China. These are bilingual social and natural scientists who often work as consultants for international environmental NGOs. Nature conservation has been the key realm in which debates about the indigenous have emerged in China.

While indigenous advocacy organizations have not been allowed to work in China, a number of environmental organizations have, for the most part, been welcomed. For decades before they came to China, these organizations were pressured by indigenous groups to respect their rights (Dowie 2009). A growing number of environmental NGO representatives extol the environmental virtues of indigenous people, who are seen to possess a unique relationship to the environment and a broad repertoire of place-specific knowledge, as well as following sustainable livelihood practices. They are said to deserve special rights. NGOs and social activists increasingly link environmental issues with social justice. By 1995, hundreds of domestic NGOs worked at the intersection between social justice and the environment in countries such as India (Princen and Finger 1994). China only registered its first environmental NGO in that year. This organization, like most other domestic environmental NGOs in China, does not work from a social justice perspective. Rather, the vast majority of environmental NGOs working in China see themselves as watchdogs against local peoples, working to ensure their compliance with state environmental laws (Weller 2006). Few Chinese environmental NGOs oppose or attempt to transform state mandates in order to create better livelihoods for rural peoples, such as by advocating for "indigenous rights."

An Indigenous Space

Lacking a strong social justice advocacy community, China reveals how the emergence of indigeneity is contingent on political activism and a willingness of people to self-identify as indigenous. The politics of indigeneity do not always result in a series of accomplishments. Rather, there are a diverse set of hopes, strategies, and practices. Even apparent victories may be contradictory, precarious, and fleeting. I use the concept of an "indigenous space" to examine a process of creating a figurative place from which one can articulate claims, gain recognition, and form alliances. It is a zone in which new ways of understanding indigenous peoples are created through the diverse efforts of numerous people.

In this view, indigenous space is never a finished project. Particular spaces are created on multiple fronts, by differently motivated actors, and always encounter diverse forms of resistance (Nelson 1999). They are created not only by the indigenous social movements that scholars of indigenous politics have focused on, but also by a wide range of practices, such as the production of indigenous media like films and radio programs (Ginsburg 1991), work in the

UN (Muehlebach 2001), indigenous art (Myers 1991), and mapmaking (Hale 2006; Chapin and Threlkeld 2001).

Efforts to establish an indigenous space do not merely attempt to gain traction for the category of indigenous people on a tabula rasa. More importantly, this work must reconfigure or displace previous understandings and rearrange social hierarchies. Westerners often expect indigenous groups to live in oral-based societies rather than writing-based ones, regarding orality as a sign of authentic indigeneity. From the perspective of most Chinese, however, oral-based societies are stigmatized not only for illiteracy but also for what was viewed as an inability to create a written language. Scholar-activists working to advocate an indigenous space for ethnic minority groups in China that lack written languages must therefore negotiate particularly strong stigmas against such groups.

In Latin America and elsewhere, the most vocal advocates for indigenous rights have often been indigenous themselves. However, almost all of the scholar-activists I observed in Yunnan identified as members of the ethnic majority, Han Chinese. Recently, some of China's ethnic minority intellectuals argued against characterizations of their own groups as backward, and documented their groups' contributions to the nation-state (Litzinger 2000), often through tactics of ethnic boosterism (McCarthy 2009). However, very few, if any, ethnic minority leaders are attempting to position their own groups as indigenous, or are working in alliance with other groups.

To create an indigenous space, Chinese scholar-activists must operate at different social and cultural registers. They need to articulate with the terminology of transnational NGOs, organizations such as the World Bank, and also find a language and style of argument that works in China. For example, many proponents of "indigenous knowledge" in China translate the term into Chinese as *chuangtong zhishi*, or, more literally, "traditional knowledge," an important difference. The term "chuangtong" is the same used in the expression "Chinese traditional culture," a concept that, in the last decade, has begun to gain acceptance, state support, and positive connotations.

Finding Chinese terms for "indigenous" has been difficult, however, as many of the terms that might be applied to indigenous people, such as *tuzhu ren* or *tu ren*, have historically been derisive. As one scholar-activist rhetorically asked, "Can we take these old terms for our aborigines, like *tuzhu ren*, and queer them? Can we make them positive?"

While I did not find the kinds of social performances of indigeneity that I expected, I found that some scholar-activists were working to displace these negative images and create a more positive framing of the indigenous in China. They are doing this work almost exclusively as it relates to environmentalism, and in particular, in the context of global concerns about habitat loss and species extinctions. Thus, concerns over the loss of biological diversity offer political leverage for the creation of cultural diversity, but in a different way than has previously existed in China. Although many in the West still imagine China as the "land of the blue ants," where individual, ethnic, and gender differences are submerged in a sea of blue Mao suits, today's China is quite different. Since the 1980s, the state has been actively involved in promoting what some have called "communist multiculturalism" (McCarthy 2009). The Chinese state has officially designated 56 different ethnic minority groups, each defined by its language, clothing, beliefs, and position on a social scale of evolution, with Han Chinese regarded as the highest. In part to boost tourism, since the 1990s, federal and provincial governments have promoted certain "colorful ethnic customs," especially those oriented toward performance for spectators, such as festivals and ethnic dress. Yet the claims that these ethnic minorities deserve greater and distinct political rights have often been dismissed.

Thus, scholar-activists who attempted to promote the concept of indigenous space had to negotiate two dominant frames for understanding ethnicity: a notion that associates cultural alterity with a lack of knowledge, sophistication, and development, and another notion that perceives ethnic difference as either choreographed entertainment or evidence of anachronistic survivals from a largely bygone era. Consequently, scholars in China who attempted to find indigenous knowledge had to negotiate a social context that made its reception and acceptance quite challenging. Although many Chinese scholars found the notion of "indigenous knowledge" quite different from the earlier mandate to uplift the ethnic minorities through instruction, a few actively used it to argue that some ethnic minority villagers did possess recognizable forms of knowledge (Agrawal 1995). In the following section, I examine how two Chinese scholar-activists attempted to document indigenous knowledge and foster an indigenous space through their engagement with environmentalism.

Making an Indigenous Space

Yu Xiaogang and Xu Jianchu have contributed in different ways to creating and reworking the meaning of indigenous space in China. Yu Xiaogang, a scholar-activist, was often referred to as a *dingzi*, literally "a nail," or one who dares to challenge authority, although usually by invoking established rules. He was part of the early wave of interest in indigenous knowledge in Yunnan, and was among the most confrontational of those advocating for indigenous space. In 1992, he received a fellowship for graduate study in Thailand, and returned to Yunnan to document "indigenous knowledge" for his master's thesis. When he carried out his project, he crossed the bounds of Chinese scholarly decorum in two ways. First, he visited several villages that had been relocated after a nature reserve was created on their old lands. He then frankly discussed the hardships the move had brought these villagers, living in a new place with poor land and little access to clean water. The nature reserve staff, mainly ethnic Han and Dai (an ethnic minority group nearly equal in number and social power to the Han in this region), described these villagers as mere peasants: ignorant, ecologically destructive, and backward ethnic minorities. Foreign organizations and agencies saw "indigenous" villagers as having far more rights than "peasants," and Yu Xiaogang argued that the villagers were sophisticated indigenous people, and he documented what he called "women's indigenous knowledge." He did not merely declare them to be indigenous, but provided detailed information on their use of medicinal plants, which many regarded as evidence of indigenous knowledge. His thesis, written in English, circulated among international NGOs, and was often cited as evidence that there was indigenous knowledge in Yunnan.

Second, Yu Xiaogang broke the bounds of decorum by openly challenging his Chinese peers and the staff of the nature reserve. At the time, many of them held to the dominant discourse about nature and society in Yunnan, which posited that ignorant, desperate, and scientifically illiterate peasants threatened the natural world. Yu's unprecedented advocacy for an indigenous space, however, made him a unique interlocutor. In one case, he carried out a "survey" among members of the nature reserve staff, the same agency that had relocated the villagers. He asked staff members a series of leading questions regarding indigenous knowledge, such as why they didn't share more power and decision-making with the indigenous people who fell within their purview. For example, staff were asked to respond "agree," "disagree," "neutral," or "don't know" to the statement "The indigenous people have rich knowledge of local ecosystems and environment; they even developed some strategies for sustainable use of natural resources." After making several such statements, Yu asked about, and more or less promoted, the idea that managers should "co-manage" with "indigenous peoples." Even though Yu was not necessarily able to convince the managers of his convictions, his actions were unconventional and path-breaking. His survey challenged managers' deeply ingrained belief that they protected the forest from local peoples, and Yu's thesis was also cited by international environmental NGOs as evidence that nature reserve managers needed reform and retraining, as they did not respect local peoples. In 2000, the head of the World Wildlife Fund recommended Yu to me as one of the few scholars looking at issues of gender and indigeneity in China. Few NGOs, however, directly advocated for indigenous knowledge and rights in China at this time, likely perceiving such actions as too confrontational.

Yet, after he moved to northwest Yunnan in the late 1990s, Yu Xiaogang began to shift away from promoting indigenous knowledge and rights per se. At first, he saw his efforts as laying the groundwork for indigenous rights in China. Yet, over time, he reasoned that if he advocated only for those particular groups in China that could be easily represented as indigenous people, his mandate would be severely restricted. Instead, Yu tried to further his own vision of a just society in which rural communities were respected and allowed to participate in major decisions affecting their lives. Later, he largely abandoned efforts to promote indigenous space per se, but still drew from alternative social models that were gaining traction internationally, such as models of communities that work together cooperatively and manage resources in surrounding areas. Yet he did not just repeat international truisms. Rather, his actions always took place within Yunnan's specific sociohistorical context. The vocabulary of Mao Zedong, Karl Marx, and *minzu* (a Chinese term related to understandings of ethnicity) were interwoven in his talk and his tactics.

Although many Americans imagine that Chinese activists must work surreptitiously, Yu Xiaogang aimed for visibility. Also in contrast to many outsiders' expectations, he did not consider himself "anti-state." Rather, he appealed to multiple levels of the state apparatus, following strategies akin to what other scholars have described as "righteous resistance" (O'Brien and Li 2006). Most surprisingly to many domestic and foreign observers, he framed himself as more loyal than government leaders to the unachieved aims of Mao Zedong. He frequently described himself as a "fundamentalist communist," a neologism only possible in today's China. The term refers to what he saw as an increasing religious fundamentalism in the United States and elsewhere, and also to the general lack of interest in communism that he saw in China, as the government party turned increasingly toward the market economy. This self-description was genuine, and it also served to deflect criticism that he was too confrontational. In contrast, one of Yu's colleagues, Xu Jianchu, has also worked on indigenous space, but he has done so strictly as an academic and consultant, never as a social organizer.

Xu Jianchu began his career by earning a degree in horticulture at a well-known university in northern China and went on to become one of China's key "cultural brokers" for issues about indigeneity and the environment. Like Yu, he earned a graduate degree in an English-speaking university outside of China, which increased his cultural fluency in international environmentalism. He has played a major role in shifting environmental discourses in Yunnan toward a much more serious consideration of indigenous knowledge. He did this in two main ways. First, he challenged the standard discourses on slash-and-burn agriculture in Yunnan. Second, he extended the notion of "sacred lands" into Yunnan. In pursuit of these two projects, he published widely in Chinese and English, worked as a consultant for many international environmental organizations operating in Yunnan, established his own highly successful NGO (the Center for Biodiversity and Indigenous Knowledge), and orchestrated a number of major international conferences in Yunnan that showcased his work and that of his students and like-minded colleagues. It was through these means that his work gained influence, as it intersected with changing international conceptions of the environment and rural people.

One of the most important ways in which Xu challenged dominant environmental discourses was by questioning understandings of slash-and-burn agriculture. By the late 1980s, Chinese officials and foreign environmentalists regarded slash-and-burn as the biggest threat to the world's tropical forests, including Yunnan's. This sense of threat animated a large number of national and international projects during the 1980s and 1990s that ended up relocating villages, especially those situated in areas deemed ecologically valuable, such as the villages where Yu Xiaogang did research. In many places, slash-and-burn was officially banned, often without any viable alternatives being offered. Slash-and-burn was largely seen in China as a destructive activity that was carried out only by ethnic minority peoples, especially uplanders.

During the 1990s, some environmentalists became reluctant to use the term "slash-and-burn," which was now seen as an excessively derogatory term that implied the wanton destruction of the forest. Instead, as part of a changing sensibility, these environmentalists employed a more neutral term, "swidden," resurrected in 1955 by anthropologist Eilert Ekwall. The new use of an older term represented a growing sensibility among environmentalists that swidden was not necessarily damaging. This concept did not automatically gain acceptance in China, but had to be promoted through particular activities in order to gain a foothold.

Xu's research worked to dislodge existing understandings and forge new views on swidden in three main ways. First, he reexamined the grounds on which swidden was judged a major reason for soil erosion. Older research compared the erosion from slash-and-burn against the erosion in undisturbed forest plots. By this comparison, slash-and-burn seemed a major problem, and these reports were part of compelling arguments for banning swidden. Xu, on the other hand, was among the first in China to compare swidden with alternative land uses, such as rubber and tea plantations. This reframing had dramatic consequences, for it vividly showed that, compared to many other land uses, swidden was less rather than more erosive.

The older research implied that slash-and-burn farmers should be relocated in order to protect the forest. Once soil erosion was shown to be even greater under plantation agriculture, slash-and-burn was no longer singled out as a major threat. Instead, it was suddenly regarded by those who prioritized soils as perhaps one of the best existing uses of land. In this way, science, rather than being used to document the destructive capacity of agriculture as practiced by upland ethnic minorities, became a tool to show that it could be more sustainable than state agricultural practices.

Second, Xu showed that slash-and-burn was advocated by state officials, rather than persisting despite their disapproval (Xu, Fox, et al. 1999). The conventional wisdom saw rural upland farmers in spatial and historical isolation. In this view, peasants defied state laws that were designed to protect forests. In contrast, Xu showed that the state had historically encouraged farmers to practice slash-and-burn as part of government-led campaigns for grain production, and to convert forests to state-run rubber and tea plantations. Thus, state-led forces had played a far more important role in deforestation than upland farmers' self-initiated actions. Xu's work shows how the emergence of indigenous space meant that identities were transformed in relationship to each other, not only between ethnic minority groups and Han, but also between villagers and the state.

Third, Xu appealed to environmentalists by showing that swidden agriculture led to much higher rates of biodiversity than alternative forms of land use (Xu, Li, et al. 1995). Where social science reports merely remarked that swidden plots appeared diverse, or listed the species found in them, Xu performed statistical analysis to document diversity. He created charts showing the "richness," "biodiversity indices," and "evenness" of species in ways that were recognized by the field of conservation biology. His charts showed that rates of diversity in older fallow compared favorably to that in "natural forest." In this way, he made quantitative arguments in behalf of Yunnan's upland farmers. His arguments succeeded in promoting the concept of indigenous space among his Chinese peers, at least to the point that they accepted his figures. As well, some expatriate conservationists also appreciated Xu's work. As one expatriate conservationist working in Yunnan told me, reading anthropological accounts might be interesting, but they were unsystematic and unreliable. However, he said, charts like those made by Xu Jianchu were clear and convincing.

Xu's savvy rethinking of the question of swidden was just one of the ways in which he influenced the emergence of the concept of indigenous space in Yunnan. He also advocated for one of the key terms of indigenous space, "sacred lands." Internationally, the concept of sacred lands gained influence in the 1980s, during the initial linkages between environmental and indigeneity.

Sacred lands were seen by some environmentalists as an example of what they called "indigenous conservation." In particular, sacred forests were offered as a prominent example of such conservation, with cases in India and throughout Africa. Xu was instrumental in spreading this notion in China, working closely with his mentor, ethnobotanist Pei Shengji. In 1985, Pei was the first to connect China with the international discourse about sacred lands when he published an article in English about the sacred forests of southern Yunnan (Pei 1985). He conducted botanical transects of these sacred forests, and, with his quantitative scientific methodologies, he convinced many domestic and international conservationists that such human-fostered forests were of great ecological importance (Hathaway 2010).

It would be difficult to overemphasize how much the language of indigenous knowledge and environmental sustainability in Pei and Xu's reports represented a strikingly new way of understanding ethnic minorities. By the early 1990s, there were some positive discussions about valley-dwelling ethnic minority groups, such as the Bai and the Dai. It is not surprising to see groups like the Dai portrayed favorably: they have a fair degree of political clout, a sizable population, and a reputation for commercial savvy; they are recognized as Buddhists (Buddhism is one of China's five officially permitted religions); and they possess their own written language. Opening up a positive space for upland ethnic minorities, who have little political power, are seen as superstitious animists, and lack their own written language, however, faces much greater challenges. But Xu's narratives and images, supplemented by studies in the idiom and mode of internationally recognized science, worked to displace ethnocentric notions and provide new interpretations. His work has energized a new generation of scholars in China who have begun to search for and document Yunnan's "indigenous forest management systems" (Xu 1991). Such systems are now presumed to be complex and culturally distinct institutions for sustainable natural resource use, worthy of study and even promotion. Until recently, such systems did not even exist in China as objects of investigation.

In many ways, this new understanding and the emergence of indigenous space in Yunnan were made possible by Pei and Xu's ongoing work. At recent conferences, such as the Cultures and Biodiversity International Congress, dozens of their colleagues articulated their passionate interest in sacred lands, indigenous knowledge, and the links between cultural and biological diversity (Xu 2000). Their research has opened up a vision of the social landscape entirely different from that of a decade ago. When the world's wealthiest environmental NGO, the Nature Conservancy (TNC), started a project in Yunnan in 1999, its efforts shaped and were shaped by the indigenous space that was created in part by Yu, Xu, and Pei. TNC used images and studies to push for an indigenous space in Yunnan for ethnic Tibetans. On the other hand, although TNC hired Xu Jianchu to look for sacred lands in Yunnan, the organization challenged his claim that such lands existed in a particular place, the region around Lao Jun Mountain. Whereas Xu described this place as inhabited by indigenes with indigenous knowledge and sacred lands, TNC staff represented its inhabitants as villagers who lacked environmental knowledge, and believed that their understanding and actions would be improved through the promotion of ecotourism. Thus TNC has pushed for some ethnic groups to be recognized as indigenous, but not others. In general, because environmental NGOs often aim at curtailing the use of natural resources, not extending the rights of rural groups to use their lands as they see fit, it is rare for them to fully embrace the concept of indigenous space. When environmental NGOs work in places like South America, their recognition of indigenous identities and respect for indigenous rights are often closely scrutinized, but in China such recognition is not established and is politically contentious. Thus, advocates for indigenous space in China have not always agreed with each other. Indeed, they have at times worked at cross-purposes.

While some scholarship on ethnicity in China has now replaced the term "ethnic minority" with "indigenous people," the example of TNC shows that indigenous advocates will sponsor only some groups and not others. Such categorical slights of hand, substituting the term "indigenous" for "ethnic minority," do not, in themselves, create particularly significant social and material changes in Yunnan. Nor can indigeneity be made relevant in China simply by coming up with a Chinese translation of the term, or having the concept imposed by powerful global players such as the World Bank. Rather, indigenous space is created cumulatively by all of these ongoing activities. Notable among them are efforts to extend the concept in specific contexts, such as environmental efforts in Yunnan, where scholar-activists actively attempt to dislodge and transform previously dominant notions and relations.

TNC's actions, as well as those of Xu Jianchu, point to the kinds of work that are necessary in China, often in the form of reports and maps of sacred lands, to position certain groups as indigenous. TNC's actions also indicate that, from the organization's perspective, certain ethnic groups, such as ethnic Tibetans, are much more compelling as indigenous subjects than others, and can become a means of gaining outside funds and attention.

Rather than taking a legalist approach to indigeneity that defines categorical criteria, the notion of indigenous space illuminates the shifting strategies, plans, and approaches used by different groups in relationship to each other. Although much work on indigeneity has revealed the struggles and the successes of the indigenous movement, such scholarship has shown us less about what kind of background work makes the concept of the indigenous salient in moments of political and environmental tension, such as those found in China. As well, viewing indigeneity in this light can provide insight into the ongoing difficulties of maintaining and expanding an indigenous space, even in countries where notions of the indigenous are well established.

Conclusion

Much scholarship on indigeneity focuses on grassroots efforts, but what we see in China is the incipient background work required for the concept of indigenous space to gain traction in the first place. In China, questions of indigeneity have not been seen as a separate issue of human rights or indigenous rights, but have mostly revolved around environmental issues. This is possible partly because, throughout the world, indigenous groups are claiming a new negotiating position in relationship to global environmentalism, and are disrupting existing frameworks. It remains to be seen how such demands will be accommodated, but challenges like this accumulate; they are not one-off events. Given changes in the international political climate, these indigenous demands, situated within a larger field of advocacy and support, cannot be easily dismissed. They demonstrate that the conceptions and status of indigenous people, and their relationships to the environment and the state, are not stable but are continually being redefined. In turn, governments, multilaterals, and conservation organizations are facing new and heightened scrutiny when their work involves people who regard themselves, and are regarded by others, as indigenous peoples.

Yet, while many common understandings of indigeneity assume that indigenous identities are based on self-definition, in Yunnan most rural citizens—including those who would most likely be deemed indigenous by outsiders—have never heard of the concept. Instead, the major interlocutors are urban-based activists. And the trajectories of these scholar-activists are shifting. Yu Xiaogang has largely stopped pushing for indigenous space. He does not want to confine his arguments to the few groups in China, like Tibetans, who have a chance of being recognized as indigenous, while disregarding the many others. On the other hand, he still draws on strategies inspired by indigenous advocacy, such as showing that villagers possess much ecological knowledge. Xu Jianchu, however, is still committed to pushing for indigenous space, and continues to search Yunnan for examples of indigenous knowledge that are compelling to domestic and international scholars and funders. In order to gain an audience within Yunnan, Xu used botanical reports to challenge conventional understandings of swidden agriculture. Moreover, his studies of sacred forests used methods from conservation biology to create authoritative accounts that argue on statistical grounds that these forests are places of ecological importance. These accounts also work against Chinese notions of sacred forests as places of feudalistic superstition, and of ethnic minorities as ignorant and environmentally threatening. In other words, in Yunnan, efforts to create an indigenous space rely on the production of scientific botanical reports, maps, and other analyses. In turn, these efforts have had a significant impact on the politics and practices of conservation in China. Indigenous space is changing these politics not only in China but throughout the world, albeit in very different ways. Elsewhere, conservation efforts have also been strongly influenced by new understandings that connect biological and cultural diversity, that use the threat of extinction to inspire new moral frameworks, strategies of development, and future worlds.

ACKNOWLEDGMENT

Elements of this chapter have been previously published in my 2010 article "The Emergence of Indigeneity: Public Intellectuals and an Indigenous Space in Southwest China" (*Cultural Anthropology* 25[2]:301–333).

NOTES

1. The word or concept "activist" has very different connotations in Chinese and English. The English term "activist" usually connotes those who work for social change, often against forces of institutional power such as corporations or the government. On the other hand, the most common Chinese terms for "activist" are *huodong zhe, huodong yuan,* and *jijifenzi,* which refer to those who actively promote state-mandated campaigns or state-approved projects. Historically, then, there is no real equivalent to the North American understanding of "activist." Indeed, in China, those who worked against state protocols were often labeled as "counterrevolutionaries" (*fan geming*). Yet we are now seeing the rise in China of some activists in the North American sense, especially in the fields of HIV/AIDS and environmentalism. An older Chinese expression that I more often heard to describe those activists who challenged authority was *dingzi,* "nail." The term is ambivalent, and can be used in reproach, appreciation, or both.

2. Such understandings, of course, did not start with Mao. Since the beginning of the twentieth century, intellectual Chinese understandings of the nation's cultural and historical legacy have often been framed in deeply critical ways. China's susceptibility to invasion from European and Japanese troops and its depiction as the "sick man of Asia" fostered such critical introspection. Conversely, the valorization of "Chinese traditional culture" only really gained serious momentum in the 1980s, as China was undergoing what some described as "culture fever" (Rofel 1994).

3. UNESCO designates certain locations as World Heritage sites only after a lengthy application process, during which the many nominated sites compete against one another. China's first World Heritage site was named in 1987, and now the government proudly proclaims China's 40 such sites, the third most of all the countries of the world. Some of the places, such as the Temple of Confucius, memorialize origins, whereas the places that focus on living cultural distinction often utilize the discourse of extinction.

4. Folk culture, not explicitly framed as the domain of "ethnic minorities," is often understood as consisting in "unique," "colorful," or "local" customs. In the last decade a number of glossy journals, somewhat pricey and designed for an urban audience, have highlighted stories about distinctive folk cultures, often said to be disappearing. A more recent trend has been grassroots and organized efforts to "preserve" such cultural dynamics, even including "do it yourself" DVD sets, with motivational messages and techniques to preserve these forms, often described as 民间杰出技艺迫 (outstanding folk skills), usually associated with arts, crafts, song, or dance. One DVD comes with a guide with the following chapters:

不能拒绝的神圣使命
(The sacred mission that can't be rejected)

责任重于泰山
(Our responsibilities are heavier than Taishan Mountain)

抢救民间文化是我们永远的职责
(It is always our responsibility to rescue folk culture)

同历史对话.以利于建设未来
(Dialogue with history to better construct the future)

我们亟需反省与反思
(What we urgently need to rethink and reflect on)

民间文艺及其开发与保护
(Development and conservation of folk culture)

文化自觉与抢救
(Cultural consciousness and cultural rescue)

抢救"国宝"
(Rescue "national treasures")

不可再做历史的罪人
(Don't become criminals of history again)

抢救民间文化刻不容缓
(Rescuing folk culture can't wait another second)

Overall, the sense is one of great urgency, with a mission and sense of responsibility, a connection to nationalism and Chinese patrimony, and a deep regret over past actions, such as the "destroy the four olds" campaign, which turned ordinary Chinese into "criminals of history."

5. As it has elsewhere around the world, language extinction in China has garnered attention and concern, including from the government. With state blessing, a number of researchers are working to stem linguistic extinction. Their efforts focus on creating writing systems for oral languages and writing dictionaries and grammar guides. Some of these linguists argue that the texts themselves will keep the language safe, and stress that the government cannot require people to continue to speak their languages. Therefore, unlike projects aimed at oral language revitalization found throughout the Americas, these efforts are primarily oriented toward the production of texts (see Nelson 1999). Yet, on the other hand, even this form of support is rarely present more broadly, as many officials believe that ethnic minority languages are ill equipped to confront modernity. Therefore, many programs for multilingual education have been dropped in favor of monolingual education in Mandarin Chinese (Bilik 2005).

REFERENCES

Agrawal, Arun
    1995 Indigenous and Scientific Knowledge: Some Critical Comments. Indigenous Knowledge and Development Monitor 3(3–4):413–439.

Bilik, Naran
    2005 Schooling Civil Society among China's Minorities. *In* Manufacturing Citizenship: Education and Nationalism in Europe, South Asia and China. Véronique Bénéi, ed. Pp. 210–235. New York: Routledge.

Brosius, J. Peter
    1997 Endangered Forest, Endangered People: Environmentalist Representations of Indigenous Knowledge. Human Ecology 25(1):47–69.

Chapin, Mac, and Bill Threlkeld
    2001 Indigenous Landscapes: A Study in Ethnocartography. Alexandria, VA: Center for the Support of Native Lands.

Conklin, Beth A., and Laura R. Graham
    1995 The Shifting Middle Ground: Amazonian Indians and Eco-politics. American Anthropologist 97(4):1–17.

Dowie, Mark
    2009 Conservation Refugees: The Hundred-Year Conflict between Global Conservation and Native Peoples. Cambridge, MA: MIT Press.

Ginsburg, Faye
1991 Indigenous Media: Faustian Contact or Global Village? Cultural Anthropology 6(1):92–112.

Gray, Andrew, Robert Harrison Barnes, and Benedict Kingsbury
1995 Indigenous Peoples of Asia. Ann Arbor, MI: Association for Asian
Studies.

Gray, Patty A.
2005 The Predicament of Chukotka's Indigenous Movement: Post-Soviet
Activism in the Russian Far North. Cambridge: Cambridge University Press.

Hale, Charles R.
2006 Más Que un Indio (More Than an Indian): Racial Ambivalence and
Neoliberal Multiculturalism in Guatemala. Santa Fe, NM: School of American Research Press.

Hathaway, Michael
2010 The Emergence of Indigeneity: Public Intellectuals and an Indigenous
Space in Southwest China. Cultural Anthropology 25:301–333.

Hodgson, Dorothy L.
2002 Precarious Alliances: The Cultural Politics and Structural Predicaments
of the Indigenous Rights Movement in Tanzania. American Anthropologist
104(4):1086–1097.

Kearney, M.
1995 The Local and the Global: The Anthropology of Globalization and
Transnationalism. Annual Review of Anthropology 24:547–565.

Li, Tania Murray
2000 Articulating Indigenous Identity in Indonesia: Resource Politics and
the Tribal Slot. Comparative Studies in Society and History 42(1):149–179.

Litzinger, Ralph A.
2000 Other Chinas: The Yao and the Politics of National Belonging. Durham, NC: Duke University Press.

McCarthy, Susan K.
2009 Communist Multiculturalism: Ethnic Revival in Southwest China.
Seattle: University of Washington Press.

Miller, Bruce Granville
2003 Invisible Indigenes: The Politics of Non-recognition. Lincoln: University of Nebraska Press.

Mueggler, Erik
    2002 Dancing Fools: Politics of Culture and Place in a "Traditional National-
    ity Festival." Modern China 28(1):3–38.

Muehlebach, Andrea
    2001 "Making Place" at the United Nations: Indigenous Cultural Politics at
    the U.N. Working Group on Indigenous Populations. Cultural Anthropology
    16(3):415–448.

Myers, Fred
    1991 Representing Culture: The Production of Discourse(s) for Aboriginal
    Acrylic Paintings. Cultural Anthropology 6(1):26–62.

Nelson, Diane M.
    1999 A Finger in the Wound: Body Politics in Quincentennial Guatemala.
    Berkeley: University of California Press.

Niezen, Ronald
    2003 The Origins of Indigenism: Human Rights and the Politics of Identity.
    Berkeley: University of California Press.

O'Brien, Kevin J., and Lianjiang Li
    2006 Rightful Resistance in Rural China. Cambridge: Cambridge University
    Press.

Pei, Shengji
    1985 Some Effects of the Dai People's Cultural Beliefs and Practices upon
    the Environment of Xishuangbanna, Yunnan, China. In Cultural Values
    and Human Ecology in Southeast China. Karl L. Hutterer, A. Terry Rambo,
    and George Lovelace, eds. Pp. 321–339. Ann Arbor: University of Michigan
    Press.

Princen, Thomas, and Matthais Finger
    1994 Environmental NGOs in World Politics. New York: Routledge.

Rofel, Lisa B.
    1994 "Yearnings": Televisual Love and Melodramatic Politics in Contempo-
    rary China. American Ethnologist 21(4):700–722.

Schein, Louisa
    2000 Minority Rules: The Miao and the Feminine in China's Cultural Poli-
    tics. Durham, NC: Duke University Press.

Sponsel, Leslie E.
    1995 Indigenous Peoples and the Future of Amazonia: An Ecological An-
    thropology of an Endangered World. Tucson: University of Arizona Press.

Sylvain, Renée
  2002 "Land, Water, and Truth": San Identity and Global Indigenism. American Anthropologist 104(4):1074–1085.

Trouillot, Michel-Rolph
  1991 Anthropology and the "Savage Slot." *In* Recapturing Anthropology: Working in the Present. Richard Fox, ed. Pp. 17– 44. Santa Fe, NM: School of American Research Press.

Tsing, Anna Lowenhaupt
  1999 Becoming a Tribal Elder, and Other Green Development Fantasies. *In* Transforming the Indonesian Uplands: Marginality, Power and Productions. Tania Murray Li, ed. Pp. 157–200. Amsterdam: Harwood Academic Publishers.

Turner, Terence
  1991 Representing, Resisting, Rethinking: Historical Transformation of Kayapo Culture and Anthropological Consciousness. *In* Colonial Situations: Essays on the Contextualization of Ethnographic Knowledge. History of Anthropology 7. George. W. Stocking, ed. Pp. 285–313. Madison: University of Wisconsin Press.

Weller, Robert P.
  2006 Discovering Nature: Globalization and Environmental Culture in China and Taiwan. Cambridge: Cambridge University Press.

Xu Jianchu
  1991 Study on Indigenous Agroecosystems in a Hani Community. MS thesis, Kunming Institute of Botany, Kunming, China.

Xu, Jianchu, ed.
  2000 Links between Cultures and Biodiversity: Proceedings of Cultures and Biodiversity International Congress 2000, July 20–30. Yunnan, China: Yunnan Press of Sciences and Technology.

Xu, Jianchu, Yanhui Li, Shengji Pei, Sanyang Chen, and Kanlin Wang
  1995 Swidden-Fallow Succession in the Mengsong Area of Xishuangbanna, Yunnan Province, China. *In* Regional Study on Biodiversity: Concepts, Frameworks, and Methods. Shengji Pei and Percy Sajise, eds. Pp. 183–196. Kunming, China: Yunnan University Press.

Xu, Jianchu, Jefferson Fox, Xing Lu, Nancy Podger, Stephanie Leisz, and Xihui Ai
  1999 Effects of Swidden Cultivation, State Policies, and Customary Institutions of Land Cover in a Hani Village, Yunnan, China. Mountain Research and Development 19:123–132.

Part 3. Red-Listed Languages

6. LAST WORDS, FINAL THOUGHTS

COLLATERAL EXTINCTIONS IN MALISEET LANGUAGE DEATH

Bernard C. Perley

Face-to-Face with Oblivion

In the summer of 1994 a meeting was organized to gather support for a Maliseet language immersion program for the Tobique First Nation community in New Brunswick, Canada. The meeting took place in the native language classroom at the reservation elementary school. In attendance were the native language teacher, the organizer of the meeting, the Head Start teachers, several mothers of children attending the school, and me. We were all trying to look comfortable in the child-sized desks. Once we had settled in, the organizer began the meeting by arguing that the reservation needed a Maliseet language immersion program. The participants were discussing the merits of the proposal while the organizer distributed photocopies of articles on language endangerment in Canada. As everyone scanned the photocopies, the organizer called our attention to the appendix of one article. She pointed out a chart that listed three categories. First listed were the aboriginal languages spoken in Canada. Second was the number of speakers speaking each language in comparison to the population of that community. The last column indicated the state of the language on a scale of "viable," "endangered," "on the verge of extinction," and "extinct." As a group we all flipped the pages until we found the listing for Maliseet. There, in cold black-and-white text, Maliseet was listed as "on the verge of extinction." Everyone in the room was silent as we all contemplated what "extinction" meant to each of us. The organizer allowed that moment of silence to continue until it turned into group discomfort. When she broke the silence, she reiterated her belief that the only way to avoid Maliseet language extinction was to support a Maliseet language immersion program for the school and for the community. One additional article predicted that, within two decades, only three aboriginal languages would be spoken in Canada. Maliseet was not one of them. That morning, we all had to come to grips with the prospect of Maliseet language extinction within the next two decades. The realization that the Maliseet language could become extinct within our lifetimes was not only discouraging but also suggested collateral extinctions that would undermine Maliseet cultural survival. In light of such dire predictions, it is difficult to find any positive or encouraging news.

Native Anthropology

I play a dual role in this story, as both native and anthropologist. Both roles are partial perspectives, in both senses of the word. I have partial (or incomplete) knowledge of both anthropology and Maliseet worlds. I also have a partial (or biased) perspective on how the two worlds intersect. I do not have the luxury of being an "outsider" in the Maliseet community, so I must negotiate community expectations and judgments. Some members of the community have very negative attitudes toward anthropologists, and I had to allay their suspicions. On the other hand, the anthropology community has a long history of condemning anthropologists who have "gone native." The most celebrated example in American anthropology is Frank Hamilton Cushing. His "adoption" of Zuñi language and customs, and his advocacy for the Zuñi community, offered a cautionary tale for subsequent American anthropologists working in Native American communities (Thomas 2000; Hinsley 1983; Jones, Jr. 1970; Deloria 1998; Gronewold 1972). For British social anthropologists, the dangers of "going native" were intimated by Evans-Pritchard in his account of fieldwork. However, he famously argued, "If an anthropologist is a sensitive person it could hardly be otherwise" (1976:245). Although he acknowledges the difficulties of "going native," he argues for its necessity. Postcolonial theorists would continue to critique the social science paradigm of objectivity as a continuation of colonial domination, but, disturbingly, also saw "going native" as a source of contamination (Torgovnick 1990; Ashcroft 1998:115). Furthermore, "going native" has been explored and critiqued as a broad sociocultural phenomenon (Deloria 1998; Huhndorf 2001; Li 2006). More recently the drama of "going native" was played out in James Cameron's blockbuster movie *Avatar* (2009). "Going native" was not and is not a problem for me, because I am "native." My problem was "going anthropologist"! Early in my professional training I had to resolve what seemed to be mutually antagonistic and exclusive communities; the dichotomy of worlds occasions the practice of epistemic slippage. In another essay, I write,

> As I reflect upon the phenomenon of anthropologists "going native" and its implications for the politics of representation, I realize that "going native" is a process of *knowing* Others. I refer to that process as "epistemic slippage." Specifically stated, epistemic slippage is our ability as anthropologists to slip from one episteme into another; that is, moving in and out of different systems of knowledge. (Perley n.d.)

As an anthropologist I was able to determine the variety of factors that contribute to Maliseet language extinction. However, as a member of the community I asked myself a critical question: "Am I part of the problem or part of the solution?" I vowed to become part of the solution. I have asserted that I am not a partial member of either community—anthropology or Maliseet. Rather, I am a full member of both communities (Perley 2009). My dual membership is the key reason I am able to envision an engaged anthropology that does not merely diagnose the relative viability of the Maliseet language. It also allows me to use my anthropology skills to find the best ways to benefit the community and revitalize the Maliseet language.

The Maliseet language is one of the many languages that have been predicted to become extinct within twenty years. It is an Eastern Algonquin language primarily spoken in the Canadian province of New Brunswick, and also in the state of Maine. My early research indicated that the twenty-year prediction could come true if no community effort was made to maintain and revitalize the Maliseet language. My position as a native anthropologist has driven me to find the most significant reason why the Maliseet language is falling out of use and to develop projects to revitalize it. In this chapter, I present the cultural and political context that continues to undermine Maliseet language vitality, the precarious state of the Maliseet language and concordant collateral extinctions, and finally, one of the potential solutions I have developed for revitalizing the Maliseet language.

Human Rights and Colonial Wrongs

On September 13, 2007, the United Nations adopted the Declaration on the Rights of Indigenous Peoples. Four countries voted "no": the United States, Australia, New Zealand, and Canada (United Nations Permanent Forum on Indigenous Issues 2010a).[1] The Canadian Broadcasting Corporation reported that

Canada's UN ambassador, John McNee, said Canada had "significant concerns" over the declaration's wording on provisions addressing lands and resources, as well another article calling on all states to obtain prior informed consent with indigenous groups before enacting new laws or administrative measures. (Canadian Broadcasting Corporation News 2007)

At issue was Article 26, stating,

1. Indigenous peoples have the right to the lands, territories and resources which they have traditionally owned, occupied or otherwise used or acquired.

2. Indigenous peoples have the right to own, use, develop and control the lands, territories and resources that they possess by reason of traditional ownership or other traditional occupation or use, as well as those which they have otherwise acquired.

3. States shall give legal recognition and protection to these lands, territories and resources. Such recognition shall be conducted with due respect to the customs, traditions and land tenure systems of the indigenous peoples concerned. (United Nations 2007:10)

CBC News reported that McNee found the provision "'overly broad, unclear and capable of a wide variety of interpretations' that could lead to the reopening of previously settled land claims and existing treaties" (Canadian Broadcasting Corporation News 2007).

While the CBC's report suggests that land, resources, and prior occupation by indigenous peoples are at the core of the controversy, I argue that, as the United Nations Declaration on the Rights of Indigenous Peoples stipulates, language, culture, and self-determination are inextricably integrated in land, resources, and occupation. The declaration states that the General Assembly is

convinced that the control by indigenous peoples over developments affecting them and their lands, territories and resources will enable them to maintain and strengthen their institutions, cultures and traditions, and to promote their development in accordance with their aspirations and needs. (United Nations 2007:2)

More importantly, Article 13 stipulates,

Indigenous peoples have the right to revitalize, use, develop and transmit to future generations their histories, languages, oral traditions, philosophies, writing systems and literatures, and to designate and retain their own names for communities, places and persons. (United Nations 2007:7)

These are laudable proclamations, but can words turn into deeds? Is it enough to grandly declare indigenous rights, "*affirming* . . . that all peoples contribute to the diversity and richness of civilizations and cultures, which constitute the common heritage of humankind" (United Nations 2007:2)? In Canada, we may never know. The Canadian government will not agree to a nonbinding declaration intended to protect the human rights of Canada's aboriginal peoples, because the government refuses to reopen "previously settled land claims and existing treaties." What appeared to be a hopeful sign for the self-determination of Canada's aboriginal peoples may only be the continuing legacy of colonial policies toward them, the perpetuation of colonial wrongs. But despite the Canadian government's self-interested indifference to the projected extinctions of aboriginal cultures, landscapes, religions, and languages, the indigenous peoples in Canada will continue to fight for their cultural survival.

Language Disintegration and the Prospects of Extinction

Language scholars have estimated that by the end of this century, half of the world's languages will have become extinct. Some scholars suggest that is a conservative estimate. More liberal estimates suggest a figure as high as 70 percent. As stated above, over the last two decades linguists, language scholars, and endangered language advocates have made this human crisis part of the public imagination, and they are working desperately to preserve, maintain, and revitalize many of the world's endangered languages. The challenge of their task is to make tangible an intangible property in order to garner support for their efforts. To this end, language scholars and advocates have argued that the loss of a language is a loss of "diversity" of distinct cultures (Grenoble and Whaley 1998; Nettle and Romaine 2000; Crystal 2000; Hinton and Hale 2001; Mithun 2004), of "human knowledge" (Crystal 2000; Wilson 2005; Harrison 2007), and of "identity" (Crystal 2000; Wilson 2005; Baldwin and Olds 2007). These arguments echo similar statements in the Declaration on the Rights of Indigenous Peoples. However, much more important is the loss of social relations. Language scholars and advocates are constrained by the rhetoric of their respective disciplines as well as their respective institutions. In all the works cited above, "language" is parsed into separate cultural categories or specific speech events. This dis-integration continues to contribute to the demise of the most important aspect of language; namely, the integrity of social relations maintained through exchanges of lived experience. The noted linguist K. David Harrison argues that we must recognize that the extinction of a language is also an extinction of ideas: "When ideas become extinct, we all grow poorer" (2007:vii). Harrison argues that "ideas" and "knowledge" come from experience in the world, and that knowledge grows from the exchange of experiences between members of a speech community over long periods. Therefore, questions of identity, culture, oral tradition, landscape, spirituality, and many more dis-integrated aspects of our human experience are constantly re-integrated through aboriginal social relations as mediated by distinctive aboriginal languages. The immediate challenge for members of communities whose languages are considered endangered is to maintain and revitalize the integration of all the facets of human knowledge that language mediates. For the community of Tobique First Nation, the immediate danger is the extinction of the Maliseet language. As I argued above, and as the United Nations has declared, and as many language scholars and advocates state, the extinction of the Maliseet language precipitates the loss of distinct Maliseet culture, knowledge, and identity. These losses are inextricably linked to indigenous human rights to land, resources, and prior occupation.

Maliseet Language Death and Collateral Extinctions

I began this essay with an ethnographic vignette from my fieldwork at Tobique First Nation in New Brunswick, Canada. The prospect of Maliseet language death within the next two decades is alarming. Scholars have identified many factors that contribute to language death, and my early research was focused on identifying which factors were contributing to the demise of Maliseet. Most of them are similar to those identified by Annette Schmidt (1990) as the loci of language attrition and subsequent language death—namely, assimilatory pressures from educational institutions, the hegemonic role of the mass media in further promoting English and other state languages, and missionization. Equally important are massive social change and the metamorphosis of speech communities, including social transformations such as mixed marriage, emigration, relocation, and—particularly relevant—the cessation of native language use in particular speech genres in particular speech events and domains. Many speech genres, in a variety of contexts, are less and less often mediated by the Maliseet language: political speeches at the tribal governmental offices, health consultations at the clinic, elementary class instruction at the reservation school, and, perhaps most devastating, everyday pleasantries between members of the community. Everyday social interactions in all these domains are conducted in English, not Maliseet. This cessation of Maliseet language use, the gradual dis-integration of the Maliseet language from Maliseet lives, has deleterious effects. Members of the community who acquired English as a first language have little difficulty conducting everyday social relations in English. However, those who acquired Maliseet as a first language and continue to use Maliseet in everyday interactions are experiencing a growing silence, as more Maliseet speakers pass away. One told me, "It's getting lonely." Although there are more than fifteen hundred residents on the reservation, the speaker recognized the difference between social relations in English and Maliseet. I have heard such comments as "it's funnier in Maliseet," or "the prayer means more in Maliseet." As I mentioned, I am a member of the community, and I confess that the most difficult comment for me to receive was, "Gee, Bern, it's too bad you don't understand Maliseet." I was saddened that I was unable to experience Maliseet worlds.

Why don't I understand Maliseet? My first language was Maliseet. As mentioned above, there are many reasons that indigenous languages become endangered. Schooling is one of them. When I started first grade, I did not speak English. The one memory I have of first grade is sitting in the front seat of the school bus. The bus driver, sitting in his seat, was twisting around to talk to me. His mouth was moving and sounds were coming out of his mouth, but I did not understand anything he said. It was a shock to realize that everything I knew and everything I had experienced in my life up to that point was rendered meaningless and irrelevant. Being a solitary Maliseet child in a largely white elementary school made me alien enough. But having my entire worldview muted and rendered meaningless made me silent and vulnerable. My mother decided that day that I would learn English: "If my son is to survive out there he'll have to master English."[2] That was not an isolated sentiment. Many Maliseet children were going through the same existential crisis I experienced. One elder told me, "The priests and the nuns said that our language was the language of the Devil. We were reprimanded in front of the class when we were caught speaking our language. So, when I had children, I vowed that they would not suffer the same kind of stigma that I did. So I spoke to them only in English." Just in these examples I have identified a number of language domains that contributed to Maliseet language endangerment: education, religion, knowledge, everyday social interactions, and identity. One by one, the connections between the Maliseet language and social relations were severed and silenced. The cumulative silencing of Maliseet relations has propelled the Maliseet language toward collateral extinctions.

Not only is the Maliseet language becoming extinct, but so too are Maliseet knowledge, religion, and identity. Compounding these collateral extinctions is the dis-integrated analysis of the language by scholars. The disintegration of language in all its socially relevant interconnectedness compounds the problem of language extinction rather than solving it. Analytical particularism, the linguistic practice of reducing spoken language to formal structural components while paying little attention to why those components are relevant to speakers, is a useful heuristic practice to track language contraction in a particular community. The neat categories articulated in the UN Declaration on the Rights of Indigenous Peoples, as well as by language scholars and advocates, may render a complex problem more comprehensible. However, I fear that the parts take on greater significance than the whole. As analysts and advocates, we may begin to fetishize language (Harpham 2002) and its particulars rather than examining how the whole is greater than the sum of all the parts. Some of those parts are Maliseet place names for places in the Maliseet landscape, a class conducted in Maliseet on the life cycle of salmon, and a story of why trees lose their leaves. However, they will be only dis-integrated individual pieces if we don't have a clear understanding of the whole. I propose that we reintegrate the pieces, in order to stave off collateral extinctions. We have already lost Maliseet place names. We have lost evidence of landscape transformations that were described in our oral traditions. We have already lost much esoteric medical knowledge. Now we are losing the ability to conduct everyday social relations in the Maliseet language. Those collateral extinctions need not be forever. As I stated above, we need to reintegrate all these facets of Maliseet experience so that we can continue to experience Maliseet worlds. What might such a reintegration look like? It could look something like *Apotamkon* (The Water Monster).

Apotamkon and the Reintegration of Maliseet Worlds

Apotamkon is an ethnographic graphic novel that I created in order to reintegrate Maliseet oral traditional stories into the Maliseet landscape, religion, and language. The traditional version of the story describes Apotamkon as a giant, serpent-like water monster that resides in the rivers at Tobique First Nation. However, I use the name Apotamkon to describe the monstrous beaver that threatened the Maliseet people. The graphic novel retells the traditional story better known as "The Tobique Rock," where Apotamkon becomes the giant beaver, Beechwood Dam, and corporations selling bottled water. Most important, my retelling incorporates contemporary concerns with lessons learned from what I call Maliseet "deep time" (Perley 2003). Too often the traditional stories are dismissed as "just so stories," with little relevance to contemporary issues and concerns. *Apotamkon* utilizes the framing devices and intertextual capabilities of graphic novels to reinforce the most important lesson—that we still have a lot to learn from our ancient storytellers. Many of the traditional stories describe how the world came into existence and how it was shaped into its present configuration for the good of the Maliseet people. *Apotamkon* continues that tradition by cautioning us to be ever vigilant to protect our environment for the common good. The graphic novel echoes the UN Declaration on the Rights of Indigenous Peoples in promoting and protecting indigenous rights—in this case, Maliseet tangible and intangible properties—but it also advocates protecting our shared common resources—in this case, water—for humankind. More important, *Apotamkon* reintegrates several social domains mediated by language: religion, landscape, oral traditions, and identity. In turn, the conversations across these particular domains promote a better integrated mediation of Maliseet social relations, as Maliseet community members share their experience of this story with others. These social exchanges help strengthen Maliseet social relations, and that is when the whole truly becomes greater than the sum of the parts. A closer examination of the integration of parts in *Apotamkon* will prepare the ground for explaining the significance of the most important piece for reintegration—the Maliseet language.

"Reading" Apotamkon

If all the frames were removed from each of the twenty pages of the graphic novel and placed side by side, you would see a continuous 360° landscape of the (quasi-)mythic Tobique Rock. Superimposed upon the landscape of Maliseet "deep time" are rough-edged frames for telling the story in mythic time. As the story unfolds, the graphic technique shifts from ink-wash drawings of quasi-traditional symbolic and abstract Maliseet forms to ink-wash renderings of the landscape. The overall effect is less controlled and more wash-like than the finely delineated pen-and-ink drawings. Instead of restricting the frames to single pages, I extended them to two-page spreads to invoke the larger expanses of mythic time. The storyteller's words are given in an informal, italic font. Interruptions to the story appear in a nonitalic, formal font. In addition, the frames of such interruptions are smaller, more regular, and fit on a single page. These single-page frames highlight the regulatory and abbreviated timeframes of the present. The storyteller's responses to these interruptions are still in an informal font, but they are not italic. As soon as he returns to his narrative, however, the informal italic font reappears, and the frame expands across two pages. The mythic time narrative culminates in a single, very large, mythic frame (figures 6.1 and 6.2). Immediately, another interruption returns to formal and constrained fonts and frames. The drawing style becomes much more controlled and the story is focused on the lessons for the present. Finally, the story concludes with the beginning of the story as told in Maliseet. This graphic novel is my exploration of reintegrated retellings of ancient stories as integral lessons for the present.

The graphic novel format allows me to take advantage of a contemporary form of storytelling while utilizing the multiple interpretive significations of graphic images (carefully rendered line drawings, informal line drawings, and graphite background drawings), texts (narrative as well as dialogue), frames (precise frames, rough-edged frames, and no frames), and renderings of space and time (360° landscapes of deep time, two-page renderings of mythic time, and small-frame renderings of the present). Most significantly, the graphic devices are only visual mediators of Maliseet lived experiences. The landscape that is graphically rendered is the landscape around and of the reservation. The Tobique Rock is a real presence in the memories of many members of the community, although dams constructed in the 1950s have submerged it under water, and it can no longer be seen. The story is a familiar one, shared by many members of the community. So, ideally, *Apotamkon* is meant to be experienced, not just read by one person to another. *Apotamkon,* however, is incomplete. It will be complete when, with the assistance of a fluent Maliseet speaker, the story is told in Maliseet. Although it has been told first in English, *Apotamkon* reintegrates the Maliseet landscape, Maliseet oral traditions, and Maliseet lived experiences, and it will soon integrate the Maliseet language. And, if my goal is achieved, *Apotamkon* will be *told* to others, not *read* to others, in Maliseet. The reintegration of Maliseet language into the daily events of Maliseet lives will reaffirm the integrity of Maliseet tangible and intangible properties, but most importantly, it will revitalize Maliseet social relations.

Last Words, Final Thoughts

The noted linguist K. David Harrison has identified the greatest loss to humankind when a language becomes extinct as the extinction of ideas. When the last speaker of a language utters his or her last words, so too does she or he utter last thoughts. In his book *When Languages Die,* Harrison

> explores only a tiny fraction of the vast knowledge that will soon be lost, an accretion of many centuries of human thinking about time, seasons, sea creatures, reindeer, flowers, mathematics, landscapes, myths, music, infinity, cyclicity, the unknown, and the everyday. By demonstrating the beauty, complexity, and the underlying logic of these knowledge systems, I hope to motivate more people—speakers, language-lovers, and scientists alike—to work harder to ensure their survival. (2007:viii)

Harrison, other language scholars and advocates, and the United Nations are all working to increase public awareness that many of the world's languages are on the verge of extinction. But while the United Nations seeks to institute dec-

Figure 6.2. *Apotamkon*, p. 13.

Figure 6.1. *Apotamkon*, p. 12.

larations protecting indigenous systems of knowledge and cultural practices as basic human rights, and while Harrison and others seek to preserve the wealth of human ideas, nations such as Australia, New Zealand, the United States, and Canada are willing to deny indigenous rights, thereby perpetuating colonial wrongs. The Maliseet of Tobique First Nation, and all aboriginal peoples in Canada, cannot wait for the Canadian government to accept a nonbinding UN declaration on indigenous rights. The government has made its stance perfectly clear—there will be no "reinterpretation" of the status quo. Aboriginal peoples cannot wait for intransigent neocolonial bureaucrats to change their minds. Nor can they hope that a nonbinding UN declaration will turn words into deeds. Aboriginal peoples can work with language scholars and advocates. But it is up to the community members to reintegrate all the pieces and ensure that indigenous worlds will survive, in all their beauty, complexity, and wealth of knowledge.

Earlier, I asked what reintegration might look like, and as an answer I presented my graphic novel and its goal of reintegrating Maliseet landscapes, oral tradition, lived experience, and language. This project is the result of my full membership in both anthropology and Maliseet communities, and it is just one example of the opportunities native anthropology has to offer for an engaged anthropology. I now ask, what might reintegration sound like? In English it sounds like "A loooong time ago." In Maliseet, it sounds like "*Awskomi piiiiiihce . . .*"

NOTES

1. Australia formally supported the declaration on April 3, 2009, and New Zealand did so on April 19, 2010 (United Nations Permanent Forum on Indigenous Issues 2010b, items dated April 8, 2009, and April 19, 2010; see also New Zealand Ministry of Foreign Affairs and Trade 2010). On December 16, 2010, the United States announced that it endorsed the United Nations Declaration on the Rights on Indigenous Peoples (United Nations Permanent Forum on Indigenous Issues 2010b, item dated December 16, 2010). And on March 3, 2010, on behalf of Canada's prime minister, Steven Harper, Governor General Michaëlle Jean stated before the House of Commons, "*We are a country with an Aboriginal heritage. A growing number of states have given qualified recognition to the United Nations Declaration on the Rights of Indigenous Peoples. Our Government will take steps to endorse this aspirational document in a manner fully consistent with Canada's Constitution and laws*" (Government of Canada 2010). However, Todd Russell, critic for aboriginal affairs in the opposition Liberal Party's shadow cabinet, accused the government of breaking this promise: "The Attorney General admitted this week that the government remains opposed to the adoption of the UN Dec-

laration on the Rights of Indigenous Peoples—breaking their commitment from only a few months ago . . . It is shameful and disappointing that the government would mislead Canada's First Nations, Métis and Inuit peoples on such an important matter" (Liberal Party of Canada 2010).

2. See Perley 2009 for additional consequences of this early experience of language alienation and the resultant planning of Maliseet language ideologies.

REFERENCES

Ashcroft, Bill, with Gareth Griffiths and Helen Tiffin
1998 Post-colonial Studies: The Key Concepts. London: Routledge.

Baldwin, Daryl, and Julie Olds
2007 Miami Indian Language and Cultural Research at Miami University. *In* Beyond Red Power: American Indian Politics and Activism since 1900. Daniel M. Cobb and Loretta Fowler, eds. Pp. 280–290. Santa Fe, NM: School for Advanced Research Press.

Canadian Broadcasting Corporation News
2007 Canada Votes "No" as UN Native Rights Declaration Passes. http://www.cbc.ca/canada/story/2007/09/13/canada-indigenous.html, accessed October 12, 2008.

Crystal, David
2000 Language Death. Cambridge: Cambridge University Press.

Deloria, Philip J.
1998 Playing Indian. New Haven: Yale University Press.

Evans-Pritchard, E. E.
1976 Appendix 4: Some Reminiscences and Reflections on Fieldwork. *In* Witchcraft, Oracles, and Magic among the Azande. Pp. 240–254. Oxford: Clarendon.

Government of Canada
2010 Speech from the Throne. March 3. http://www.speech.gc.ca/eng/media.asp?id=1388, accessed September 2, 2010.

Grenoble, Lenore A., and Lindsay J. Whaley, eds.
1998 Endangered Languages: Current Issues and Future Prospects. Cambridge: Cambridge University Press.

Gronewold, Sylvia
1972 Did Frank Hamilton Cushing Go Native? *In* Crossing Cultural Boundaries: The Anthropological Experience. Solon T. Kindell and James B. Watson, eds. Pp. 33–50. San Francisco: Chandler.

Harpham, Geoffrey Galt
2002 Language Alone: The Critical Fetish of Modernity. New York: Routledge.

Harrison, K. David
2007 When Languages Die: The Extinction of the World's Languages and the
Erosion of Human Knowledge. New York: Oxford University Press.

Hinsley, Curtis
1983 Ethnographic Charisma and the Scientific Routine: Cushing and
Fewkes in the American Southwest, 1879–1893. In Observers Observed:
Essays on Ethnographic Fieldwork. George W. Stocking, Jr., ed. Pp. 53–69.
Madison: University of Wisconsin Press.

Hinton, Leanne, and Ken Hale, eds.
2001 The Green Book of Language Revitalization in Practice. London: Academic Press.

Huhndorf, Shari M.
2001 Going Native: Indians in the American Cultural Imagination. Ithaca,
NY: Cornell University Press.

Jones, Oakah L., Jr.
1970 Introduction. In My Adventures in Zuñi. By Frank H. Cushing. Pp.
5–11. Palo Alto, CA: American West.

Li, Victor
2006 The Neo-primitivist Turn: Critical Reflections on Alterity, Culture, and
Modernity. Toronto: University of Toronto Press.

Liberal Party of Canada
2010 Harper Government Breaks Throne Speech Commitment on Indigenous Rights. June 2. http://www.liberal.ca/newsroom/news-release/harper-
government-breaks-throne-speech-commitment-on-indigenous-rights/,
accessed September 2, 2010.

Mithun, Marianne
2004 The Value of Linguistic Diversity: Viewing Other Worlds through Native American Indian Languages. In A Companion to Linguistic Anthropology. Alessandro Duranti, ed. Pp. 121–140. Malden, MA: Blackwell.

Nettle, Daniel, and Suzanne Romaine
2000 Vanishing Voices: The Extinction of the World's Languages. New York:
Oxford University Press.

New Zealand Ministry of Foreign Affairs and Trade
   2010 Statement by Hon. Dr. Pita Sharples, Minister of Maori Affairs: An-
   nouncement of New Zealand's Support for the Declaration on the Rights of
   Indigenous Peoples. http://www.mfat.govt.nz/Media-and-publications/
   Media/MFAT-speeches/2010/0-19-April-2010.php, accessed September 2,
   2010.

Perley, Bernard C.
   2003 Language, Culture and Landscape: Protecting Aboriginal Deep Time
   for Tomorrow. Paper presented at "Protecting the Cultural and Natural
   Heritage in the Western Hemisphere: Lessons from the Past; Looking to the
   Future," Harvard University, Cambridge, MA, December 5–7. http://projects
   .gsd.harvard.edu/heritage/program.htm, accessed October 20, 2008.
   2009 Contingencies of Emergence: Planning Maliseet Language Ideologies.
   In Native American Language Ideologies: Language Beliefs, Practices, and
   Struggles in Indian Country. Paul Kroskrity and Margaret Field, eds. Pp.
   255–270. Tucson: University of Arizona Press.
   N.d. Epistemic Slippage and the Perils of Native Anthropology. *In* An-
   thropology and the Politics of Representation. Gabriela Vargas-Cetina, ed.
   Tuscaloosa: University of Alabama Press.

Schmidt, Annette
   1990 The Loss of Australia's Aboriginal Language Heritage. The Institute
   Report Series. Canberra: Aboriginal Studies Press.

Thomas, David Hurst
   2000 Skull Wars: Kennewick Man, Archaeology, and the Battle for Native
   American Identity. New York: Basic Books.

Torgovnick, Marianna
   1990 Gone Primitive: Savage Intellects, Modern Lives. Chicago: University
   of Chicago Press.

United Nations
   2007 United Nations Declaration on the Rights of Indigenous Peoples.
   http://un.org/esa/socdev/unpfii/documents/DRIPS_en.pdf, accessed September 20, 2007.

United Nations Permanent Forum on Indigenous Issues
   2010a United Nations Declaration on the Rights of Indigenous Peoples, Adopted by the General Assembly 13 September 2007. http://www.un.org/esa/
   socdev/unpfii/en/declaration.html, accessed September 2, 2010.
   2010b News and Events. http://www.un.org/esa/socdev/unpfii/en/news
   .html, accessed September 2, 2010.

Wilson, Waziyatawin Angela
    2005 Defying Colonization through Language Survival. *In* For Indig-
    enous Eyes Only: A Decolonization Handbook. Waziyatawin Angela
    Wilson and Michael Yellow Bird, eds. Pp. 109–126. Santa Fe, NM: School
    of American Research Press.

7. DYING YOUNG

PIDGINS, CREOLES, AND OTHER CONTACT LANGUAGES AS ENDANGERED LANGUAGES

Paul B. Garrett

The fact that a large and steadily growing number of the world's languages are "endangered" has received increasing attention in recent years, in academic and professional forums as well as in the mass media. In general treatments of the topic, both scholarly and popular, a few case studies or anecdotes are typically used to lend an element of human interest and local color to the global statistics, which, though striking in and of themselves, are inevitably rather abstract. The languages chosen for this purpose tend to be relatively obscure and "exotic" ones: languages spoken by small communities in remote locales, sure to be very little known, if not entirely unknown, to the vast majority of readers, listeners, or viewers. The choice of such obscure languages is in many ways an effective strategy, particularly for the mass media, and particularly for those forms of media that rely heavily on visual images, such as magazines, television, and websites. Portrayals of "traditional" peoples and their ways of life emphasize the poignancy and human tragedy of language loss by linking it to loss in other domains: for example, loss of culturally distinctive worldviews, ancient cosmologies and ritual practices, ingenious subsistence strategies that enable humans to survive in "harsh" physical environments, and intimate understandings of local ecologies and biodiversity.

Even in those treatments of language endangerment in which a particular language and its speakers are the main focus, some sort of universalistic perspective is almost always taken. One prominent theme that often emerges is that the presumed reduction of "diversity" in various domains, such as those mentioned above, is an unfolding tragedy in and of itself, whether seen from a broad humanistic perspective or a somewhat narrower scholarly perspective (linguistic or otherwise). Another common theme is that the endangerment of languages is diagnostic of grave dangers facing not just those who speak them, but all of humanity, or all of planet Earth[1]—and by implication, the reader, listener, or viewer and his or her own way of life. Taken together, these themes suggest both a fundamental equality and a profound unity among the world's languages. It is presented as axiomatic that all human languages are equally valuable—not in practical or utilitarian terms, but in a more abstract or philosophical sense—and that each has its own unique place within a richly diverse but integrated whole that is ineffably greater than the sum of its parts. The "death" of any language, however "small" and obscure, is therefore an inestimable and irreversible loss; and the endangerment of any language is a matter worthy of everyone's attention and concern.

Or is it? My purpose here is not to critique either scholarly or popular discourses on the issue of language endangerment; there is little need, for insightful critiques of these discourses seem to be proliferating almost as rapidly as the discourses themselves. My purpose, rather, is to consider the exclusion (or, at least, the omission) of a particular category of languages from these discourses.[2] Only very rarely are contact languages—those languages, commonly referred to as pidgins and creoles, that are historically known to have taken form in situations of contact among speakers of two or more previously existing languages (Garrett 2004; Thomason 2001)—even mentioned, much less focused upon, in discussions of language endangerment. But several of these languages have already met their demise, some quite recently, such as Negerhollands (a Dutch-lexified creole formerly spoken in the United States Virgin Islands), Skepi Dutch Creole, and Berbice Dutch Creole (both formerly spoken in parts of Guyana). Various others are currently on (or rapidly approaching) the brink of "death," such as Michif, Mednyj Aleut, Chinook Jargon, Trinidadian French Creole, Unserdeutsch/Rabaul German Creole, and Pitcairn/Norfolk Island English Creole. Meanwhile, numerous others, though in less dire circumstances at present, have far from certain futures: St. Lucian French Creole, Dominican French Creole, the English Creoles of the Central American coast (e.g., Miskito Coast, Limón, Bay Islands), Palenquero, and dozens more. Why should the precarious state of so many languages go virtually ignored in discussions of language endangerment?[3]

In a review article dealing with three recently published books on language endangerment, Mühlhäusler comments, "It is worrying that none of the authors addresses the massive loss of Pidgins, Creoles, and other impure contact languages [sic], the loss of jargons, cants and argots or indeed the loss of so-called dialects. We need to beware of the narrowly focussed discursive construction of endangered languages" (2003:243–244). Beyond the volumes reviewed by Mühlhäusler, it seems particularly odd that endangered contact languages should go unmentioned in recent books by prominent scholars that deal comprehensively with the full range of language-contact phenomena (e.g., Thomason 2001 and Winford 2003; also Mufwene 2001, which highlights the *development* of contact languages, even those that are now endangered or defunct, but the *endangerment* of various noncontact languages, such as Native American languages and African languages in New World contexts[4]). In these and other such works, it may be noted in passing that specific contact languages among those considered in the chapters devoted to pidgins, creoles, etc., are endangered. Yet these same languages are *not* mentioned in the chapters or sections devoted specifically to language endangerment; the focus shifts instead to more typically cited examples of endangered languages, such as Native American languages. This seems to reflect the general tendency for research on contact languages to emphasize their origins and development, but to stop short of following their trajectories through to obsolescence and death. Exceptions do exist, but are few in number: they include Sabino's (1994, 1996) work on Negerhollands, Kouwenberg's (2000) work on Berbice Dutch Creole, Erhart and Mühlhäusler's (2007) survey of endangered Pacific pidgins and creoles, and Hazaël-Massieux's (1999) comprehensive book-length treatment of French-lexified creoles as *langues en péril* ("languages in peril").[5]

Marginal Languages

To be sure, contact languages have always been "marginal" languages, as Reinecke (1937) put it several decades ago, and as has been repeated many times since. By and large, this remains true today, despite the efflorescence of scholarship on these languages since the 1960s, the emergence of language advocacy movements in various places where they are spoken, and the fact that a few among them (such as Haitian, Tok Pisin, Papiamentu, and Seselwa) have been accorded significant degrees of official recognition and institutional support. But to what extent does this persistent marginality account for why contact languages are so rarely mentioned in discussions of language endangerment?

It would seem that endangered contact languages are in fact doubly marginalized: marginalized among the world's languages in general, and then marginalized again among endangered languages. What accounts for this secondary marginalization? How are contact languages any more "marginal" than, for example, languages of Amazonia or Papua New Guinea that have only a few hundred speakers each—languages that are little known even among linguistic scholars?

In general terms, contact languages (even when thriving) tend to have certain characteristics in common with endangered languages: on the whole, they are oral (unwritten) languages spoken by relatively small populations, lacking in political and economic power, who inhabit geopolitically remote areas such as small islands and tropical coastlines. These are among the characteristics that render the great majority of the world's 6,000-plus languages "marginal" vis-à-vis a small few "world" languages such as English, Spanish, French, and Arabic, and vis-à-vis a few hundred other languages, from Turkish to Guaraní to Catalan, that have the institutional backing of the nation-states within which they are spoken (or of a semiautonomous political unit within the nation-state, as in the case of Catalan).

But contact languages also have certain other characteristics that render them marginal even among the remaining thousands of relatively small, relatively powerless languages. These characteristics are several, but can be grouped under two related themes. One is their relative lack of historicity; the other is their perceived lack of autonomy.

Shallow Histories

The great majority of contact languages are of quite recent origin, having arisen no more than three to four centuries ago in contexts associated with European colonialism. (This is true even of some of those that do not have European lexifiers, e.g., Fanakalo and Lingala.) These languages are, as Trouillot has said of Caribbean societies, "inescapably historical" (1992:21). Their historicalness is an important part of what makes contact languages intriguing to linguistic scholars of various stripes: to a degree not possible with older languages, one can ascertain the specific time, place, and circumstances of their origins and trace their developmental trajectories up to the present day.

But the pertinent issue here is not the rich historicalness of these languages. Rather, it is their relative lack of *historicity* (in Bell's [1976] sense), which is quite another matter. Unlike historicalness, which is largely a matter of documentation and of painstaking research by specialists (usually academically trained professionals), historicity is primarily a matter of popular perception and ideology: of people's folk understandings of the relationship between a particular language and their own (or someone else's) ethnohistory. In discussing the central role of language in the formation of the "imagined communities" of modern nations, Anderson comments, "One notes the primordialness of languages, even those known to be modern. No one can give the date for the birth of any language. Each looms up imperceptibly out of a horizonless past" (1991:144–145). Their shallow time-depth makes contact languages notable exceptions to Anderson's generalization, and likewise makes them problematic for his model of nation-state formation (Garrett 2007).

Similarly, contact languages' relative lack of historicity tends to attenuate their ties to specific geographical territories and to the contemporary inhabitants of those territories. In many cases, those who speak the contact language are not indigenous to the area that they currently inhabit; they or their forebears (going back relatively few generations) were transported there as slaves or indentured servants, or else migrated there as laborers, military conscripts, traders, or settlers. Whatever the case, and whatever the juridical status of their present-day claims to the land on which they live, they cannot easily assert the kinds of primordial (not to say incontestable) ties that can be asserted by an aboriginal population that has lived in a particular place since time immemorial. Errington notes, "Aboriginality can be leveraged . . . into claims of ownership, trumping rights of access that might otherwise be claimed by and granted to encroaching 'outsiders.' These are situations in which languages can take on value if they are portrayed as organically bound up with place and culture, and as likewise under threat of encroachment" (2003:724). As Errington's observation also suggests, the ethnic identities of contact-language-speaking groups may be less coherent and less robust—in that they are less "focused" or more "diffuse" (Le Page and Tabouret-Keller 1985), and therefore likely to be more open to negotiation and contestation—than those of any aboriginal group or groups with whom the territory may be shared. (That territory may itself be the object of competing claims between autochthonous and more recently arrived groups, as is currently the case in the interiors of Guyana and Suriname.)

Another relevant consideration is suggested by Henning Andersen (1988; see also Wolfram 2002:768). Andersen distinguishes broadly between "open" and "closed" communities, referring to their degree of "openness" toward, and contact with, the outside world; and between "endocentric" and "exocentric" communities, referring to whether the community orients itself primarily to its own local, internal norms or to external norms originating elsewhere. Colonial societies, creole societies, mining towns, rapidly growing urban centers, and other settings in which contact languages typically develop all tend to be relatively "open" and/or "exocentric," for many of the same reasons outlined above. Further complicating all of these issues may be a more or less explicit awareness that the distinctive cultural traditions of contact-language-speaking groups in fact originate elsewhere, typically on a distant continent with which any direct ties have long since been severed. There may also be a more or less acute awareness that those traditions are in fact multiple and heterogeneous, and have become irreversibly intermingled and hybridized—that is to say, creolized.

Lack of Autonomy

Many of the same factors that account for contact languages' lack of historicity also account for their perceived lack of autonomy. Historically, many if not most contact languages have been regarded (even by their own speakers) as merely "broken" or "corrupted" versions of the European languages to which they are self-evidently related. The fact that this relationship is primarily lexical makes it particularly salient for those who (unlike linguistic scholars) may be wholly unaware that relationships to other languages are discernible and indeed demonstrable at other levels of analysis, such as syntax. Complicating matters is the fact that where a contact language remains in contact with its lexifier, the existence of a "creole continuum" of intermediate lects typically blurs the boundary between the two. The resultant clinal effect may tend to make contact languages problematic as symbols of distinctive group identities (ethnic, national, or otherwise)—at any rate, more problematic than languages that have no such relationship, lexical or otherwise, to the dominant languages with which they coexist (as in the case of Native American languages in contact with English, for example).

Lack of autonomy can also present certain advantages, however. Mufwene (2003) remarks on a phenomenon that may actually favor the survival of creoles and other contact languages that remain in sustained contact with their lexifiers—thus providing an answer to the broader question of why nonstandard, strongly stigmatized vernacular varieties perdure despite hegemonic, strongly normative pressures from standardized varieties "with which they have coexisted and in which their speakers acquire literacy":[6]

> An answer to this apparent puzzle may lie in the fact that, despite linguists' common claim that creoles are separate languages relative to their lexifiers, speakers of all these stigmatized vernaculars think that they speak the same language as the prestigious variety in which they are provided literacy. There is between their vernaculars and the standard variety a division of labor that creates no competition of the sort that would lead to the attrition or loss of the nonstandard and less prestigious ones. (Mufwene 2003:330–331)

Mufwene's reasoning suggests that creoles and other "stigmatized vernaculars" that are in contact with dominant languages other than their lexifiers would tend to be in greater danger. This does indeed seem to be the case. To take some well-known examples from the Caribbean region, where such situations are quite common (Snow 2000), Mufwene's observation doubtless helps account for why Dutch-lexified creoles have fared poorly in contact with English; why the English-lexified creoles of the Central American coast are being displaced by Spanish; and why French-lexified creoles have gone into decline in those islands where French has been displaced by English, such as Trinidad and Grenada, and, to a lesser extent, St. Lucia (Garrett 2000, 2005) and Dominica.[7]

This need not mean certain death for these languages; to be sure, numerous other factors must also be taken into consideration. Interestingly, such situations may actually offer certain advantages for language preservation and revitalization projects. Although these languages doubtless continue to suffer from their lack of autonomy vis-à-vis their lexifiers (many St. Lucians still speak of their creole as a kind of "broken French," for example), at the local level, the contemporary absence of the lexifier, and the contact language's self-evident difference of lexicon from the contemporary standard-official language, gives the contact language considerably greater autonomy than its counterparts in continuum situations typically have (no St. Lucian would suggest that the French-lexified creole is in any sense the "same language" as English). This greater degree of autonomy may in turn serve to heighten public awareness of language endangerment at the local level and thus create more favorable conditions for language advocacy efforts, such as those currently underway in St. Lucia (Garrett 2000); and it may tend to make the language a more potent symbol of national and/or ethnic identity (Garrett 2007).

These considerations notwithstanding, it is certainly true that on the whole, contact languages' lack of autonomy vis-à-vis their lexifiers renders them marginal. Advocates of these languages therefore face uphill battles (even more so than advocates of other endangered languages) at all levels, official and unofficial, local, regional, and international. Aside from the general reluctance of local governments and organizations to devote scarce resources to these languages, contact languages very rarely if ever attract any attention—or funding—from the handful of international organizations that are specifically devoted to the study, documentation, and preservation of endangered languages. Lack of involvement in such organizations by those who study contact languages may well be one factor in this, and is a matter well worthy of consideration.

But there are surely larger issues, an important one being the pervasive tendency to privilege strongly autonomous languages over those that self-evidently (i.e., lexically) derive from previously existing languages. Errington notes the high value placed on distinctive, "relatively culturally salient lexicons" in language endangerment discourses: languages are portrayed as repositories of knowledge and experience, and their lexicons in particular as unique referential and classificatory systems whose value "may only become apparent in the future." Lexicons are thus regarded as "relatively isolable information bases . . . which are in danger of falling out of use before they are codified, or before the uses of information they embody are discovered" (2003:724–725).[8] By this logic, the lexicon of a contact language—a pastiche of second-hand materials rather hastily assembled (either crudely or ingeniously, depending on one's ideological perspective), typically under extremely adverse circumstances only a few generations ago—can hardly be expected to hold "priceless treasures" of knowledge or wisdom that must be safeguarded for the sake of all humanity.[9]

A final consideration is that some contact languages lack autonomy even from one another. The tendency of some francophone scholars to refer collectively to all of the world's French-lexified creoles in the singular as *le créole* (which seems to be an expression of the universalistic tendencies of hegemonic French language ideology) comes to mind here, as do the guiding assumptions and practical objectives articulated by the transnational franco-creolophone movement known as the Bannzil (Hookoomsing 1993). A particular subset of the world's French-lexified creoles, those of the Lesser Antilles, presents a more focused (and therefore even more problematic) case in point. If Grenadian French Creole, for example, "dies out," will a unique "language" have died out? Or just a particular local manifestation of a larger entity, referred to by some scholars as Lesser Antillean French Creole, that remains well established and may still have good prospects for continued survival in other nearby territories?[10] In such a case, are we actually dealing with a case of language "death" at all, or merely a kind of geographical contraction? If the latter, should the endangerment of Grenadian French Creole generate as much concern (among Grenadians, Bannzil members, linguistic scholars, the general public) as the endangerment of a contact language such as Palenquero or Michif that is (arguably) more singular, both linguistically and sociohistorically?

What Is There to Lose?

A number of knotty questions have been raised, all of them relating in some way to the persistent marginality of contact languages. Given the two major reasons for that marginality that have been identified here—namely, their shallow histories and their lack of autonomy—the question that ultimately arises is: What is there to lose if a contact language "dies out"?

The most obvious answer—so obvious that it borders on trivial, perhaps—is that some part of the world's linguistic diversity is lost. A far more challenging and interesting way of posing the question is, What part? How big a part, how important a part? Here again, knotty questions quickly proliferate. When the issue is language endangerment, are all languages really to be regarded as equal? Is the "death" of a century-old pidgin such as Russenorsk as much of a loss to the world's linguistic diversity as the death of a Native American language, for example? What about cases of so-called decreolization? Do we regard the gradual loss of a creole's basilectal forms and features due to sustained contact with its lexifier in the same way that we regard the erosion of inflectional morphology in an Amazonian language as its speakers increasingly shift to Portuguese or Spanish?

These are awkward questions to which there are no simple answers. But they are surely worth asking; and it is also worth asking *why* they are awkward questions, both to pose and to answer. It could perhaps be argued that they are unfair questions, a matter of comparing apples to oranges. But such an argument is at odds with the predominantly universalistic themes of language endangerment discourses, as well as with the now broad-based consensus among linguistic scholars that the contact-induced processes involved in the genesis and development of (so-called) contact languages affect all languages to one degree or another.[11] And of course, it fails to address the broader question of why contact languages are given little if any consideration in discussions of language endangerment.

In order to address such questions meaningfully, it is necessary to consider at least briefly the metaphorical discourses through which many of them are formulated and which set the tone, if not also the terms, of ensuing discussions. Drawing analogies between endangered languages and endangered biological species is a tempting and common tactic, used carefully and thoughtfully by some, more casually and carelessly by many others. Various commentators have considered the strengths and weaknesses of such analogies, and there is no need to delve deeply into the matter here. It is worthwhile, however, to consider one particular aspect of these analogies with specific regard to endangered contact languages. Clearly, certain endangered biological species, characterized by Errington as "charismatic megafauna (pandas, whales, and so on)" (2003:724), receive far greater attention, public expressions of valorization, and funded intervention than others. Taking up the same theme, Mühlhäusler suggests that the linguistic equivalents of these charismatic megafauna are "the languages of a small number of tribal people living away from the centres of technological and social change." He goes on to suggest that a rather vapid and ineffectual "moral discourse" (comparable to pseudo-environmentalist "Greenspeak") has become "clearly dominant" over more rigorous scientific and economic discourses (2003:243).

It is abundantly clear that among the world's endangered languages, endangered contact languages are sorely lacking in the sort of "charisma" to which Errington and Mühlhäusler allude—so much so that they often seem to be virtually invisible. In terms of the endangered species metaphor, they are perhaps analogous to the snail darter: a decidedly unspectacular small fish (*Percina tanasi*) that was discovered in the Little Tennessee River in 1973, shortly before the Tellico Dam was to be built. Environmentalists and other opponents of the dam project quickly launched a campaign to have the snail darter added to the United States' official list of endangered species, and they were successful in 1975. The result was a prolonged legal battle that pitted environmentalists against the Tennessee Valley Authority, which argued that $78 million had already been spent on the dam and that the snail darter was, after all, just a small fish and should not be allowed to stand in the way of a major government works project. The case went all the way to the Supreme Court, which in 1978 decided in favor of the environmentalists (and by extension, the snail darters) by ruling that the "plain intent" of the federal government's Endangered Species Act legislation was to protect any and all endangered species, "whatever the cost." In the following year, however, the state of Tennessee's congressional delegation slipped into an appropriations bill a rider that would exempt the Tellico Dam project from the Endangered Species Act. The modified bill passed by a narrow margin, whereupon the Tennessee Valley Authority proceeded to complete the dam—thus destroying the snail darters' only known habitat.[12]

Though it should hardly be necessary, some recent critiques of language endangerment discourses remind us that the status of these languages cannot be separated from the status (social, political, economic) of their speakers. If one is willing to stretch the analogy a bit further, the case of the snail darters can serve as a similar reminder: be they biological species or human communities, small, powerless groups that appear to have nothing to offer to larger, more powerful groups—or that are perceived as presenting some obstacle or challenge to their hegemony (as when a creole language is regarded as an impediment to formal education in the standard language, or to state-sponsored "development" or "modernization" projects)—tend to be ignored and neglected at best or, at worst, forcefully and systematically eradicated. Under such circumstances, even the most vigorous, well-orchestrated, broad-based advocacy campaign may not be enough to save them.

Consideration of how the status of a contact language relates to the status of its speakers can be helpful in addressing the larger question that has been posed here: What is there to lose? With regard to pidgins, the answer may at first seem to be: Not much. After all, it is the nature of pidgins not to outlive their usefulness, as it were; they tend to be ephemeral, lasting only as long as, and developing only to the extent to which, they are needed to facilitate some fairly basic level of communication within a specific domain of activity (which may be relatively ephemeral itself). Several well-attested pidgins are known to have become "extinct" over the years, such as Russenorsk, Pidgin Basque, Chinese Pidgin Russian, and Vietnamese Pidgin French; it is unlikely that anyone mourned the passing of these language varieties any more than the dissolution or disruption of the historically specific configurations of social relations that gave rise to them and for a time sustained them. For these were always secondary, auxiliary, vehicular languages, limited in function; they were no one's primary language of identity (ethnic or otherwise), no one's primary language of sentiment and self-expression.

But as ecological perspectives emphasize, every language is part of a system that is homeostatic yet ceaselessly dynamic, complexly integrated yet fundamentally open-ended: that is, an ecology. The demise of a language is bound to have repercussions throughout such a system, and may well disrupt its equilibrium, either temporarily or permanently (in which case the system becomes subject to disintegration). As Mühlhäusler notes, "What is at risk are not individual languages but the complex ecological support system that sustains linguistic diversity" (2003:241). The quasi-Darwinian assumption (sometimes encountered in ecological and political-economic models alike) that languages are always in competition for dominance, and that an ecology of language is ultimately reducible to struggles between dominant and dominated groups, "under-emphasizes the degree of cooperation and interdependencies between languages as well as the complex layers of language ecologies with local vernaculars, regional and inter-village lingua francas and Pidgins that have long sustained structured diversity" (2003:240). Mühlhäusler goes on to explain that contact languages, in many cases, have had key roles in maintaining the equilibrium of linguistic diversity, thereby helping to perpetuate it:[13]

> Sustained linguistic diversity in the past included a range of solutions to intergroup communication, such as institutionalized multilingualism or the presence of Pidgins and lingua francas as Drechsel has demonstrated for the Mobilian language of the United States. It was the ability of speakers to communicate in other languages in particular Pidgins that helped them shield their own small languages against larger neighbouring ones. (2003:242)

As Mühlhäusler suggests, the ecological role of a pidgin may be crucial and should not be underestimated—even in cases where those who speak the language may seem to have little investment in it or allegiance to it. The decline of a pidgin may be only the beginning, or the earliest symptom, of a systemic perturbation with far-reaching consequences for linguistic, sociocultural, and biological diversity alike (Maffi 2001).

Whatever their potential ecological significance, pidgins, as noted above, are auxiliary or vehicular languages; they are no one's primary language of identity, no one's primary medium of sentiment and self-expression. Very much in contrast, creoles and other contact languages (e.g., those variously referred to in the literature as bilingual mixed languages, intertwined languages, semi-creoles, partially restructured varieties, and indigenized varieties) typically *are* quite central to the identity and self-expression of their speakers. Some contact languages (e.g., Media Lengua, Anglo-Romani, Michif) seem to have emerged for precisely these reasons: as a means of constituting and asserting, or shoring up and maintaining, a particular group's sense of its own identity, community, and ethnic or cultural distinctiveness. Similarly, in recent debates over the definition and application of the term "creole," various commentators have concluded that one of the major distinguishing features of a creole—for some, the chief defining feature—is that, unlike a pidgin, it is the primary language of some community of speakers (Jourdan 1991), and is therefore likely to be their language of ethnic identity (Garrett 2004; Jourdan 1991).[14]

So aside from ecological considerations of the kind discussed above, in the case of creoles and other nonpidgin contact languages there tends to be a keener sense, on the part of individuals as well as whole communities, that there is in fact a great deal to lose—or at the very least, there tends to be greater receptivity to such an idea, and even this can provide a crucial starting point for language advocates and activists. (However important pidgins might be shown to be in ecological terms, it is rather more difficult to imagine a successful movement to preserve or revitalize a pidgin.) When a creole or other nonpidgin contact language is endangered, there is an attendant danger that an already marginalized people, whose claims to a distinctive cultural, ethnic, and/or national identity may be tenuous at best, stands to lose a crucial foundation, a "valid symbolic substrate" (Errington 2003:730), for those very claims. To take a broader view, also at stake in each case is a unique, ever-evolving record of a unique episode of human experience. The relative brevity and recency of such episodes hardly diminish their significance.[15] For Trouillot, who calls for a "rehistoricization" of creolization as both a linguistic and a sociocultural phenomenon, "creolization is a miracle begging for analysis. Because it first occurred against all odds, between the jaws of brute and absolute power, no explanation seems to do justice to the very wonder that it happened at all" (1998:8).

What Is to Be Done?

A final, special consideration for those who study contact languages, and who care about the people who speak them, is that to allow their endangerment to go unnoticed and unremarked is to be tacitly complicit in perpetuating the kinds of structural and symbolic violence that were key factors in the emergence of so many of these languages—and that, in all too many cases, continue to confront their speakers, in one form or another, on a daily basis. At the same time, however, we must be both mindful and respectful of the fact that such conditions may be precisely what speakers are coping with as they shift (intentionally or not) from these languages to others that seem to offer (illusively or not) possibilities of social and economic advancement. The profoundly inhumane circumstances under which many contact languages emerged, and that continue to reverberate in the daily lives of their contemporary speakers, are the source of a profound ambivalence that in many communities would tend to undermine even the most heroic locally initiated efforts to "preserve" or "revitalize" these languages. Furthermore, as Labov (2008) asserts in regard to African-American Vernacular English, matters of human rights, community well-being, and social justice must take priority over matters of language vitality and diversity. This is perhaps nowhere more clear than in cases where the language (or dialect, or variety) in question is itself a product of institutionalized discrimination, inequality, oppression, and violence. It is no mere coincidence that in virtually all such cases, the language in question can be classified as a contact language; and the attendant social ills are likewise the products of contact-related social phenomena.

Ultimately, as the preceding points suggest, decisions as to whether or not efforts to protect, preserve, and/or revitalize contact languages are desirable and worthwhile must be made by individuals and communities at the local level. Scholars who are themselves native, near-native, or semispeakers of the local language, and/or members of the communities in which they carry out their research, may be especially well positioned to contribute to these decision-making processes and to lead or participate in subsequent programs of action. (It may well be the case that speakers of creoles and other contact languages are considerably better represented in academia than are speakers of most other endangered languages.) As Dorian notes, "It is extraordinarily difficult for even the most sympathetic outsiders to provide useful support for endangered small languages . . . Moral support and technical expertise, including linguistic expertise, can and should be offered, certainly, but . . . even in the event of acceptance, effective leadership can only come from inside the community" (1998:21).

As Dorian's caveat suggests, however, those of us who are only sojourners or peripheral participants in these communities surely have something to offer. Hill proposes that the most useful role that scholarly investigators of endangered languages can take is to find ways of "helping community activists to rally people in endangered-language communities to defend and reclaim their languages while simultaneously attracting resources from dominant communities" (2002:128–129). At the very least, scholarly work can help reveal and demystify the processes of symbolic domination (Bourdieu 1991), both historical and contemporary, whereby these languages have come to be endangered, thus providing local leaders with information that can help them to devise maximally pragmatic, strategically targeted programs of action. Just as important, we can expand our long-standing scholarly commitments to fostering awareness and recognition of these languages and their speakers by staking out a place for them in ongoing discussions of language endangerment. By doing so, we can help ensure that they receive the attention of government agencies, nongovernmental organizations, advocacy and activist networks, and other institutions that, in turn, may be able to help them attain some of the basic forms of recognition, legal protection, funding, and material resources that all speakers of endangered languages so vitally need in order to make unfettered, well-informed decisions, and to act on those decisions.

ACKNOWLEDGMENTS

This chapter is a revised, updated, and expanded version of an essay published in 2006 as "Contact Languages as Endangered Languages: What Is There to Lose?" in the *Journal of Pidgin and Creole Languages* 21(2):175–190. The author and editor acknowledge the kind permission of the John Benjamins Publishing Company, Amsterdam and Philadelphia.

NOTES

1. Scholarly works in which these themes are made explicit include Harmon 1996, Harrison 2007, Maffi 2001, and Nettle and Romaine 2000.

2. Nonaka 2004 draws attention to another such category: indigenous and original (as distinct from national) sign languages. These languages are also considered briefly in Harrison 2007:230–233.

3. This is not to suggest that all or most other endangered languages are specifically mentioned by name in these discussions—clearly, the sheer number of languages in question makes this infeasible. This being the case, endangered languages are often referred to collectively, in geographically defined groupings: "Amazonian languages," "aboriginal languages of Australia," "Native American languages of the Pacific Northwest," etc.

4. Mufwene 2008, another book-length work (which in many respects builds on Mufwene 2001), includes a brief case study of Gullah that considers its potentially endangered status, but focuses primarily on its persistence.

5. Also worth noting here is Ferreira's (2009) report on recent efforts to revitalize a variety of French-lexified creole spoken in Venezuela.

6. But see Wolfram and Schilling-Estes 1995, Wolfram 1997, and Guy and Zilles 2008 on the issue of dialect obsolescence and endangerment and the importance of studying it within the broader context of research on language obsolescence and endangerment.

7. Sasse's (1992) multistage model of language death (via language shift) predicts that in the final stage, the obsolescent language may persist in a more or less vestigial form, serving emblematic functions; and that a distinctive variety of the language to which speakers have shifted, more or less strongly influenced by the obsolescent language, may have emerged. See Garrett 2005 and Garrett 2003, respectively, for discussions of these phenomena in St. Lucia, where both are currently in evidence (although language obsolescence, thus far, is not in its final stages, and language death, while an imminent possibility, is not a foregone conclusion).

8. By the same token, contact languages' reputation, even among linguists, as morphologically and syntactically relatively "simple" or "simplified" (but see De-Graff 2001 for a rebuttal of such notions) surely helps account for why endangered contact languages receive far less attention than languages that exhibit unusual structural features—such as Kayardild, spoken by fewer than 100 persons on two islands off the northern coast of Australia, which is the only known language in which tense markers appear on nouns as well as on verbs, according to Bernard Comrie and Martin Haspelmath (Dorian 2002:135). For that matter, the fact that contact languages typically lack such attrition-prone features as complex agglutinative morphology may mean that their endangerment tends to go undiagnosed, as it were, by native speakers and linguistic scholars alike.

9. See Hill 2002 on "hyperbolic valorization" and "universal ownership" as prevalent themes in the rhetorics of language endangerment.

10. Further complicating these issues are recent offshoot varieties such as San Miguel Creole French, a variety spoken in Panama that is traceable to St. Lucian laborers who migrated there in the mid-nineteenth century, and that was reported in 1999 to have only three remaining speakers (Lewis 2009). See also Ferreira 2009 on a variety that is still found in Venezuela.

11. For a particularly strong and detailed statement of the latter point, see Mufwene 2008.

12. Ultimately this did not turn out to be the demise of the species after all. In the following year, snail darter populations were discovered in other sections of the river that were less drastically affected by the dam.

13. This is not always the case, however; depending on the circumstances under which they enter into local ecologies of language, contact languages can also be disruptive and detrimental to linguistic diversity (as Mühlhäusler acknowledges, and as he has demonstrated in his own work in Pacific contexts, e.g., Mühlhäusler 1996).

14. See also Mufwene 2001 (106–125), whose concern is not how creoles are to be distinguished from pidgins, but how, historically, they have been distinguished from other vernacular varieties of their lexifiers (and thus defined as creoles): "The main implicit criterion, which is embarrassing for linguistics but has not been discussed, is the ethnicity of their speakers" (2001:xiii). This observation is developed further in Mufwene 2008:93–112.

15. For that matter, contact languages may be less exceptional in these regards than is commonly thought. As Hill notes, "'languages' and their 'dialects' are, in many cases, very recent historical artifacts" (2002:128).

REFERENCES

Andersen, Henning
    1988 Center and Periphery: Adoption, Diffusion, and Spread. *In* Historical Dialectology. Jacek Fisiak, ed. Pp. 39–83. Berlin: Mouton de Gruyter.

Anderson, Benedict
    1991[1983] Imagined Communities: Reflections on the Origin and Spread of Nationalism. London: Verso.

Bell, Roger T.
    1976 Sociolinguistics: Goals, Approaches and Problems. London: Batsford.

Bourdieu, Pierre
    1991 Language and Symbolic Power. Cambridge, MA: Harvard University Press.

DeGraff, Michel
    2001 On the Origin of Creoles: A Cartesian Critique of Neo-Darwinian Linguistics. Linguistic Typology 5(2–3):213–310.

Dorian, Nancy C.
    1998 Western Language Ideologies and Small-Language Prospects. *In* Endangered Languages: Language Loss and Community Response. Lenore A. Grenoble and Lindsay J. Whaley, eds. Pp. 3–21. Cambridge: Cambridge University Press.
    2002 Commentary: Broadening the Rhetorical and Descriptive Horizons of Endangered-Language Linguistics. Journal of Linguistic Anthropology 12(2):134–140.

Erhart, Sabine, and Peter Mühlhäusler
    2007 Pidgins and Creoles in the Pacific. *In* The Vanishing Languages of the Pacific Rim. Osahito Miyaoka, Osamu Sakiyama, and Michael E. Krauss, eds. Pp. 118–143. Oxford: Oxford University Press.

Errington, Joseph
2003 Getting Language Rights: The Rhetorics of Language Endangerment
and Loss. American Anthropologist 105(4):723–732.

Ferreira, Jo-Anne S.
2009 The History and Future of Patuá in Paria: Report on Initial Language
Revitalization Efforts for French Creole in Venezuela. Journal of Pidgin and
Creole Languages 24(1):139–157.

Garrett, Paul B.
2000 "High" Kwéyòl: The Emergence of a Formal Creole Register in St. Lucia. In Language Change and Language Contact in Pidgins and Creoles. John
H. McWhorter, ed. Pp. 63–101.
2003 An "English Creole" That Isn't: On the Sociohistorical Origins and
Linguistic Classification of the Vernacular English of St. Lucia. In Contact
Englishes of the Eastern Caribbean. Michael Aceto and Jeffrey P. Williams,
eds. Pp. 155–210. Amsterdam: John Benjamins.
2004 Language Contact and Contact Languages. In A Companion to Linguistic Anthropology. Alessandro Duranti, ed. Pp. 46–72. Oxford: Blackwell.
2005 What a Language Is Good for: Language Socialization, Language Shift,
and the Persistence of Code-Specific Genres in St. Lucia. Language in Society 34(3):327–361.
2007 Say It Like You See It: Radio Broadcasting and the Mass Mediation of
Creole Nationhood in St. Lucia. Identities: Global Studies in Culture and
Power 14(1–2):135–160.

Guy, Gregory R., and Ana M. S. Zilles
2008 Endangered Language Varieties: Vernacular Speech and Linguistic
Standardization in Brazilian Portuguese. In Sustaining Linguistic Diversity:
Endangered and Minority Languages and Language Varieties. Kendall A.
King, Natalie Schilling-Estes, Lyn Fogle, Jia Jackie Lou, and Barbara Soukup,
eds. Pp. 53–66. Washington, D.C.: Georgetown University Press.

Harmon, David
1996 Losing Species, Losing Languages: Connections between Biological and
Linguistic Diversity. Southwest Journal of Linguistics 15:89–108.

Harrison, K. David
2007 When Languages Die: The Extinction of the World's Languages and the
Erosion of Human Knowledge. Oxford: Oxford University Press.

Hazaël-Massieux, Marie-Christine
1999 Les Créoles: L'Indispensable Survie. Paris: Éditions Entente.

Hill, Jane H.
    2002 "Expert Rhetorics" in Advocacy for Endangered Languages: Who Is
    Listening, and What Do They Hear? Journal of Linguistic Anthropology
    12(2):119–133.

Hookoomsing, Vinesh Y.
    1993 So Near, yet So Far: Bannzil's Pan-Creole Idealism. International Jour-
    nal of the Sociology of Language 102:27–38.

Jourdan, Christine
    1991 Pidgins and Creoles: The Blurring of Categories. Annual Review of
    Anthropology 20:187–209.

Kouwenberg, Silvia
    2000 Loss in Berbice Dutch Creole Negative Constructions. Linguistics
    38(5):889–923.

Labov, William
    2008 Unendangered Dialects, Endangered People. In Sustaining Linguistic
    Diversity: Endangered and Minority Languages and Language Varieties. Pp.
    219–238. Kendall A. King, Natalie Schilling-Estes, Lyn Fogle, Jia Jackie Lou,
    and Barbara Soukup, eds. Washington, D.C.: Georgetown University Press.

Le Page, Robert B., and Andrée Tabouret-Keller
    1985 Acts of Identity: Creole-Based Approaches to Language and Ethnicity.
    Cambridge: Cambridge University Press.

Lewis, M. Paul, ed.
    2009. Ethnologue: Languages of the World. 16th ed. Dallas, TX: SIL Interna-
    tional. http://www.ethnologue.com/, accessed March 24, 2011.

Maffi, Luisa, ed.
    2001 On Biocultural Diversity: Linking Language, Knowledge and the Envi-
    ronment. Washington, D.C.: Smithsonian Institution Press.

Mufwene, Salikoko S.
    2001 The Ecology of Language Evolution. Cambridge: Cambridge University
    Press.
    2003 Language Endangerment: What Have Pride and Prestige Got to Do
    with It? In When Languages Collide: Perspectives on Language Conflict,
    Language Competition, and Language Coexistence. Brian D. Joseph, Neil G.
    Jacobs, and Ilse Lehiste, eds. Pp. 324–345. Columbus: Ohio State University
    Press.
    2005 Créoles, Écologie Sociale, Évolution Linguistique. Paris: L'Harmattan.
    2008 Language Evolution: Contact, Competition and Change. London:
    Continuum.

Mühlhäusler, Peter
    1996 Linguistic Ecology: Language Change and Linguistic Imperialism in
    the Pacific Region. London: Routledge.
    2003 Language Endangerment and Language Revival. Journal of Sociolin-
    guistics 7(2):232–245.

Nettle, Daniel, and Suzanne Romaine
    2000 Vanishing Voices: The Extinction of the World's Languages. Oxford:
    Oxford University Press.

Nonaka, Angela M.
    2004 The Forgotten Endangered Languages: Lessons on the Importance of
    Remembering from Thailand's Ban Khor Sign Language. Language in Soci-
    ety 33(5):737–767.

Reinecke, John
    1937 Marginal Languages: A Sociological Survey of the Creole Languages
    and Trade Jargons. Ph.D. dissertation, Yale University.

Sabino, Robin
    1994 . . . They Just Fade Away: Language Death and the Loss of Phonological
    Variation. Language in Society 23(4):395–426.
    1996 A Peak at Death: Assessing Continuity and Change in an Underdocu-
    mented Language. Language Variation and Change 8(1):41–61.

Sasse, Hans-Jürgen
    1992 Theory of Language Death. In Language Death: Factual and Theoretical
    Explorations with Special Reference to East Africa. Matthias Brenzinger, ed.
    Pp. 7–30. Berlin: Mouton de Gruyter.

Snow, Peter
    2000 Caribbean Creole/Non-lexifier Contact Situations: A Provisional Sur-
    vey. Journal of Pidgin and Creole Languages 15(2):339–343.

Thomason, Sarah G.
    2001 Language Contact: An Introduction. Washington, D.C.: Georgetown
    University Press.

Trouillot, Michel-Rolph
    1992 The Caribbean Region: An Open Frontier in Anthropological Theory.
    Annual Review of Anthropology 21:19–42.
    1998 Culture on the Edges: Creolization in the Plantation Context. Planta-
    tion Society in the Americas 5(1):8–28.

Winford, Donald
    2003 An Introduction to Contact Linguistics. Oxford: Blackwell.

Wolfram, Walt
    1997 Issues in Dialect Obsolescence: An Introduction. American Speech
    72(1):3–11.
    2002 Language Death and Dying. *In* The Handbook of Language Variation
    and Change. J. K. Chambers, Peter Trudgill, and Natalie Schilling-Estes, eds.
    Pp. 764–787. Oxford: Blackwell.

Wolfram, Walt, and Natalie Schilling-Estes
    1995 Moribund Dialects and the Language Endangerment Canon: The Case
    of the Ocracoke Brogue. Language 71:696–721.

Part 4. Prehistories of an Apex Predator

8. DEMISE OF THE BET HEDGERS

A CASE STUDY OF HUMAN IMPACTS ON PAST AND PRESENT LEMURS OF MADAGASCAR

Laurie R. Godfrey and Emilienne Rasoazanabary

This chapter results from the collaborative efforts of Laurie Godfrey, a primate paleontologist, and Emilienne Rasoazanabary, a specialist on the behavior of living nonhuman primates. Both of us study the primates that live or once lived on the island of Madagascar—lemurs. In this chapter, we examine extinction, taking as our example recent extinctions on Madagascar (including the extinction of giant lemurs) and threats to the smaller-bodied lemur species that remain there today. Extinctions can be viewed in deep time, in near time, or in today's world; each view generates insights that cannot be gained from any of the others. A "deep time" perspective is usually reserved for extinctions that occurred before humans evolved, so humans cannot have been responsible. There have periodically been major mass extinctions in the past (called extinction "events" because of the unusually high number of species extinctions concentrated in relatively short periods of time), each with different but profound effects on the evolutionary history of life on Earth. Quaternary extinctions, "extinctions in near time," demand a consideration of humans as at least possible agents of extermination (MacPhee 1999). It was during the very last part (the most recent 100,000 years) of the geological period called the Quaternary (or Pleistocene and Holocene) that people began to populate many regions that had never before experienced their presence, and these regions, one after the other, suffered dramatic species loss. In many ways, such "near-time" extinctions rivaled or surpassed some of the worst mass extinctions of the distant past, and tools that have been applied to the analysis of species rarefaction in the deep past have been applied as well to late Quaternary extinctions.

From the differences in pace and pattern of species rarefaction, we can infer changing extinction processes. Quaternary extinctions are remarkable for their disproportionate representation of large-bodied ("megafaunal") species. In certain other mass extinctions, tiny sea creatures, reef-building corals, and insects were primary victims. Thus, the study of Quaternary extinctions has focused largely on why large-bodied species (the good "bet hedgers," as we will see) are most vulnerable to extinction at the hand of humans. It has also focused on why certain faunal communities, such as those on Madagascar, were hit harder than communities elsewhere. Megafaunal extinction generally entails the disproportionate disappearance of species weighing 40 kilograms (around 100 pounds) or more. On Madagascar, the body mass extinction threshold is much lower. Madagascar was one of the last places on Earth to be inhabited by humans, and megafaunal extinctions there were very recent. This makes them relatively easy to study. Many remnants of the recent presence of megafauna on the island still can be found.

By reconstructing extinction events of the recent past, we can hope to gain a depoliticized understanding of the impacts of humans (as a formidable predator and strong competitor for resources) on other species. We can hope to observe extinction without becoming enmired in the politics of income disparity, cultural differences, and cultural self-determination. Nevertheless, we cannot escape the realization that extinction has cultural consequences. Perhaps it is not coincidental that, as humans spread, the species that die off first are also the ones that contemporary societies prize.

Madagascar: A Biodiversity Hotspot

At 226,656 square miles, Madagascar (the Great Red Island) is the world's fourth largest island and one of the Earth's top five biodiversity "hotspots"—so called because together they contain very high percentages of the Earth's vascular plant species (20 percent) and vertebrate species (16 percent), while covering a minute portion (0.4 percent) of the Earth's land surface (Sodhi 2008). Madagascar is renowned as the home of the lemurs, its flagship species. Its human population is greater than 20 million today, and increasing at around 3 percent annually. Its rate of natural increase is approximately five times that of the United States.

Forest deterioration and fragmentation is occurring more rapidly today than in the past (Smith et al. 1997; G. Harper et al. 2007). The island's habitats are ecologically diverse, and include eastern rain forests, central montane forests and savannas, western dry deciduous forests, and southern gallery and xerophytic spiny forests. Today the central highlands are largely savanna, but it is unlikely that the highlands were ever densely forested. There is strong evidence that the grasses covering the interior today are not anthropogenically derived, as was recently maintained, but rather that grasses invaded Madagascar soon after the late Miocene epoch (Bond et al. 2008). Whereas there were more wooded habitats in the central highlands in the past than there are today, particularly as corridors along waterways, rampant forest habitat destruction is likely a relatively recent phenomenon tied to human population increase, and it is not responsible for a spread of grass across the island's interior.

Broadly speaking, climatic conditions on Madagascar have remained essentially modern over the past 5 million years (Wells 2003). The monsoon system that influences Madagascar's hypervariable climate today has an ancient history, as it can be tied to the rise of the Tibetan Plateau and the associated Himalayan uplift that resulted from the collision of India and southeast Asia at least 35 million years ago (Ali and Huber 2010; Ali, personal communication). Madagascar's monsoons had likely reached today's high intensity by around 8 million years ago.

This is not to imply that there were no climate fluctuations during the past 5 million years or so. Indeed, fluctuations in the climate and vegetation of Madagascar associated with the rise and fall of sea level during the late Pleistocene and Holocene have been well documented (e.g., Burney 1993; Burney, James, et al. 1997; Burney, Burney, et al. 2004; Gasse and Van Campo 1998, 2001; Wells 2003; Virah-Sawmy et al. 2009, 2010). The last glacial maximum (cool period) occurred 18–20,000 years ago, well before the arrival of humans on Madagascar. This was followed by a relatively moist phase, and then a period of aridification beginning around 3,000 years ago, again prior to the arrival of humans. No faunal extinctions are known to have been associated with any of the pre-human-colonization events. The ecogeographic body-size gradient that characterizes today's extant lemurs also characterizes the same species in the recent past, suggesting similar bioclimatic pressures and resource seasonality well into the past millennium (500 years at a minimum; Muldoon and Simons 2007).

Extinctions did occur after the arrival of humans on Madagascar, and they were wide-ranging and spectacular. There was no blitzkrieg; rather there was a prolonged yet inexorable extinction episode. That episode eliminated all endemic species of Malagasy vertebrates larger than 10 kilograms in body mass, and several that were less than 10 kilograms. The larger species included at least 17 lemurs (in five families: Daubentoniidae, Lemuridae, Megaladapidae, Palaeopropithecidae, and Archaeolemuridae), all of Madagascar's flightless "elephant birds" (*Aepyornis* and *Mullerornis* spp.), giant tortoises, a horned crocodile (*Voay*; Brochu 2007), several species of hippopotamus (Faure and Guérin 1990), and a euplerid carnivore (*Cryptoprocta spelea*; Goodman, Rasoloarison, et al. 2004). Also eliminated were two species belonging to an unusual aardvark-like genus (*Plesiorycteropus*) that has been placed in its own mammalian order (MacPhee 1994), some large volant raptors (Goodman 1994; Goodman and Rakotozafy 1995), and the southern giant jumping rat (*Hypogeomys australis,* Goodman and Rakotondravony 1996). Smaller yet, but also apparent victims of the recent extinction event, were a shrew tenrec (*Microgale macpheei,* Goodman, Vasey, et al. 2007) several additional species of rodents (Mein et al. 2010), and several species of bats, the largest of which weighed barely over 200 grams (Samonds 2007).

Today's ecosystems little resemble those of the period just prior to human colonization of the Great Red Island. For example, in the central highlands at Ampasambazimba some two thousand years ago, there were ~20 species of primates alone (including many that are still extant), along with hippopotamuses, elephant birds, giant tortoises, and so on. All of the endemic large-bodied species are extinct, and none of the smaller-bodied lemur species still lives in this region (Godfrey, Jungers, Reed, et al. 1997). In the spiny and gallery forests of the southwest, up to four species of living lemurs can be found; an additional eight species of now-extinct lemurs recently lived in this region, along with elephant birds, giant tortoises, hippos, a species of *Plesiorycteropus,* the horned crocodile, and others. The relative abundance of endemic versus introduced rodents has changed dramatically. In certain parts of Madagascar today, the forests have been invaded by introduced rats and mice; even 150 to 100 years ago, endemic rodents were far more common than they are today (e.g., compare Grandidier 1902 and Youssouf Jacky and Rasoazanabary 2008 on the abundance of endemic rodents in the gallery forests of southwestern Madagascar). Introduced dogs and wild cats may be replacing endemic carnivores in some regions as the most important terrestrial predators of living lemurs (Brockman et al. 2008).

Chronology of Extinction on Madagascar

Several lines of evidence converge to suggest that humans first settled in southwestern Madagascar around two thousand years ago or slightly earlier (Burney, Burney, et al. 2004): (1) early dates on human-modified bones of extinct species (four *Hippopotamus,* one elephant bird tibiotarsus, and one radius of the giant lemur, *Palaeopropithecus*) at several sites (Taolambiby, Lamboharana, Itampolo, and Ambolisatra) (see Burney, Burney, et al. 2004; Perez et al. 2005); (2) the first occurrence of pollens of introduced plants (notably *Cannabis*); (3) a precipitous decline in the spores of the mega-dung fungus, *Sporormiella,* a good proxy for megafaunal biomass (Burney, Robinson, et al. 2003; Burney, Burney, et al. 2004); and (4) a spike in charcoal microparticles well above prior background levels. Burney, Burney, et al. (2004) base their scenario for extinction (figure 8.1) on evidence that the megafaunal decline on Madagascar preceded habitat destruction. Table 1 shows the earliest dates associated with human-modified bones in southwest Madagascar, with their appropriate 2σ error terms; there is no indication of human presence prior to 2,300 BP. Several hundred years later (i.e., between 230 and 410 AD), sites in the southwest experienced a decline in *Sporormiella* spores, followed by an abrupt increase in charcoal microparticles (Burney, Burney, et al. 2004). This sequence of events can be identified at several sites, spreading from the southernmost locations up the west coast and into the interior. Burney, Burney, et al. (2004) interpreted it as signaling, first, human-induced decimation (without extinction) of the megafaunal populations, likely via hunting, followed by a spike in human-induced fires. The identity of the earliest settlers is unknown, although some scholars have defended an African origin (e.g., Blench 2007). Crowley's (2010) recently refined radiocarbon chronology demonstrates that very large-bodied taxa declined before large taxa, which in turn declined before small- and medium-sized species. It also documents that the declines began shortly after our posited time of arrival of humans, and well before a thousand years ago.

Definitive evidence of people arriving from Indonesia as well as from Africa comes later, in the form of direct archaeological evidence of a contemporaneous association of Bantu TIW pottery and shell-impression pottery of likely Austronesian influence. This coalescence of cultures sprang from an ancient Indian Ocean trade network that bridged southeast Asia and the east coast of Africa. The Malagasy language belongs to the Barito group of languages from southeast Kalimantan (Borneo) (see Allibert 2008 for a review), but has a secondary, predominantly Bantu, component. Genetic evidence also unequivocally supports a combined Asian and African heritage characterized by multiple migration waves, the first having occurred some time prior to 1500 years ago, and possibly as early as 3000 years ago (Hurles et al. 2005; Tofanelli et al. 2009).

The earliest definitive settlement sites on Madagascar date from between the fifth and eighth centuries AD (Dewar and Wright 1993; H. Wright and Rakotoarisoa 2003; Dewar 2003). However, by early in the second millennium AD, human settlements had spread all over Madagascar. That spread also corresponded with a period of aridification, culminating in a severe drought ~950 calibrated years BP (Virah-Sawmy et al. 2010). This climate change may have impacted the megafauna in some regions directly, and it may have impacted human populations and accelerated the decline of megafaunal populations by necessitating increased human reliance on bushmeat for survival. However,

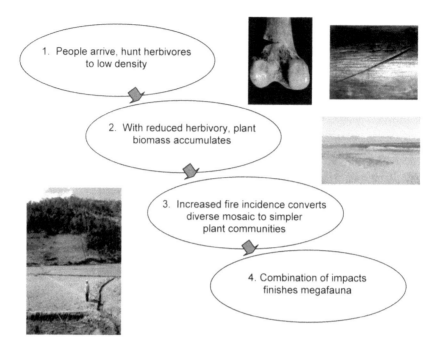

Figure 8.1. Scenario of megafaunal extinction in Madagascar.
(Based on data from Burney, Burney, et al. 2004.)

at approximately the same time, the spore count of *Sporormiella* rose again in some places, providing possible evidence of the introduction (or population expansion) of humped cattle (*Bos indicus*) from Asia (Burney, Burney, et al. 2004). The cattle may have come to Madagascar indirectly from Africa, as Austronesians also introduced humped cattle into continental Africa. There is some controversial evidence that the latter introduction occurred prior to two thousand years ago, but definitive archaeological and skeletal evidence that it had occurred by the eighth or ninth century AD (Magnavita 2006). The timing of the first introduction of cattle to Madagascar is unknown, but we know that cattle, rice, chickens, fish, turtles, and wild animals were consumed at Mahilaka, the first major port on Madagascar. This port was established in the twelfth century AD and flourished until around 1400 AD (H. Wright and Rakotoarisoa 2003).

Europeans did not discover Madagascar until the year 1500 AD, at first entirely accidentally. By the time Europeans arrived, Madagascar's megafauna were rare, and only one direct encounter is described in the European litera-

Table 1. Earliest dates on human-modified bones in southwest Madagascar (from Burney, Burney, et al. 2004).

Genus	Calibrated date (±2σ)*	Notes
Hippopotamus	60 BC–AD 130	
Hippopotamus	AD 155–415	
Hippopotamus	805 BC–AD 640	Large error term
Hippopotamus	2020 BP (not calibrated)	Redated; first date rejected.
Aepyornis	30 BC–AD 320	
Palaeopropithecus	2366–2315 BP	

*unless otherwise noted

ture—that of an elephant bird shot and killed by a startled merchant named Ruelle (Hébert 1998). This incident occurred only a decade after French explorer-naturalist Étienne de Flacourt (1658) had described elephant birds (which the Malagasy called "Vorompatra"—the bird of the Ampatres, the region of the Antandroy in southern Madagascar). Flacourt noted that his informants insisted the bird still lived in the Ampatres. Also on the basis of Malagasy accounts, Flacourt (1658) described a giant lemur (the "tretretretre," undoubtedly *Palaeopropithecus,* see Godfrey and Jungers 2003) and a ferocious "man-eating" animal (this was the hippo, although Flacourt didn't recognize it as such from its description). Hundreds of years later, European folklorists recorded the tales of the songomby, the tsy-aomby-aomby or kilopilopitsofy (the hippo), and a giant ogre that could be rendered helpless on smooth flat surfaces. The latter was undoubtedly the same beast that Flacourt had identified as the tretretretre. Hippos were known to Malagasy people in some parts of Madagascar well into the nineteenth and twentieth centuries (Godfrey 1986; Burney and Ramilisonina 1998); it is likely that they were among Madagascar's longest megafaunal survivors. It is also possible that the giant lemurs *Megaladapis* and *Archaeolemur* (as well as *Palaeopropithecus*) survived into the last half of the second millennium (Burney and Ramilisonina 1998), but the most recent radiocarbon dates confirm only the survival of *Megaladapis madagascariensis* and *Palaeopropithecus ingens* until five or six hundred years ago (Burney, Burney, et al. 2004). The subfossil site yielding these late dates is a pit cave called Ankilitelo, located in southwestern Madagascar. It contains the remains of several additional species of extinct lemur (*Archaeolemur majori, Pachylemur insignis,* and *Daubentonia robusta*). If Muldoon et al. (2009) are correct in assuming that the pit caves of the southwest sample narrow windows of time, then all five of these extinct lemur species would have been alive five hundred years ago. Radiocarbon dates testify to the survival of other giant lemurs (including *Hadropithecus stenognathus, Pachylemur insignis,* and *Mesopropithecus pithecoides*), as well as flightless birds, at least until the end of the first millennium AD (Burney, Burney, et al. 2004).

Across the island, Madagascar's human population was likely sparse until the second half of the second millennium. Prickly pear cactus was first introduced into the arid south in 1769 (Kaufmann 2001); it became an important dry-season food supplement for cattle, resulting in a dramatic increase in the carrying capacity of the southern spiny forests, and concomitant increases in the cattle and human populations. Historical records summarized by Agarwal et al. (2005) document the construction of roads, the felling of trees, the setting of fires, and overgrazing in many parts of Madagascar while the island was under French colonial control. Forest concessions were encouraged by the French, as was selective conversion of forest to plantations for cash crops. By 1900, the human population of Madagascar was estimated to be around 2.5 million; a century later, that number had grown to almost 16 million, and in 2010, it surpassed 20 million (Population Reference Bureau 2010).

The twentieth-century increase in human population brought with it an acceleration of deforestation, which today threatens thousands of additional species with extinction (G. Harper et al. 2007). The actual pace of forest loss is well documented for the period following 1953, when the extent of forest cover was established via aerial photographs. Landsat images came later (1973, 1990, and 2000). In the latter half of the twentieth century, forest cover on Madagascar decreased by 40 percent and core forest (i.e., forest located more than 1 km from nonforest habitat) decreased by almost 80 percent (G. Harper et al. 2007). All forest types suffered loss, including the humid forests (43 percent from the 1950s to ~2000), dry forests (41 percent), and spiny forests (28 percent). At a single well-documented site in western Madagascar (Andranomena), Smith et al. (1997) describe a loss of forest cover of 44 percent between 1950 and 1990. Whereas it is difficult to reconstruct the "original" (i.e., precolonization) primary forest cover of Madagascar, few researchers believe that more than 10 percent remains today (see Ingram and Dawson 2005; Hannah et al. 2008). G. Harper et al. (2007) estimate that half to two-thirds of precolonization forest cover was depleted prior to the mid-twentieth century.

The Selectivity of Extinction:
Why Do Large-Bodied Animals Disappear First?

Whereas today many small-bodied as well as some of the remaining larger-bodied endemic animals are threatened with extinction on Madagascar, it is clear that during the earlier phase of extinction (i.e., extinction in "near time"), the largest-bodied animals were the first to disappear. A shift in the ecological profile of the most vulnerable species likely reflects a change in the most important threat to species. The primary trigger of extinction or local extirpation has changed from human hunting (targeting large species) to forest fragmentation (targeting frugivores) (Godfrey and Irwin 2007).

Globally speaking, Quaternary human-mediated extinctions reveal a clear bias against large-bodied species. Much has been written about that size bias, and despite ongoing debate (for reviews, see Lyons et al. 2004; Koch and Barnosky 2006), its causes are fairly well understood. Both intrinsic and extrinsic factors come into play. Intrinsic factors that can influence the probability of extinction include such variables as r_{max} (the intrinsic rate of maximal population increase) and generation length. Extrinsic factors that can affect the probability of extinction include such variables as predation (including human hunting) and changes in food availability.

In a short paper published several years ago, Brook and Bowman (2005) address the intrinsic factors that allow size-based extinction predictions to be made; they call this the allometry of extinction. These authors make three critical observations: First, r_{max} generally scales with body mass raised to the -0.25 power. It is lower in large-bodied than small-bodied animals. Second, generation length (G) generally scales with body mass raised to the $+0.25$ power; it is longer in large-bodied animals. Third, the additional per-generation mortality required to exceed the ability of a population to replace itself even under the best of all reproductive conditions, when realized reproduction equals maximum reproduction, scales with the product of r_{max} and G, and thus is a constant when the two exponents are reciprocals. Large animals are therefore no more at risk than small animals if they experience a comparable percentage increase in mortality over the span of a generation. But any increase in annual kill rate will affect large-bodied species more than small ones, by virtue of the larger species having longer generation times. Thus, an identical annual increase in mortality will create a significant bias against large-bodied animals, while a disproportionate increase in mortality of large-bodied animals, springing from their being specifically targeted, will have an even greater negative impact. Human hunting, an extrinsic factor, creates such a biased mortality increase.

What is less well understood are the reasons that similarly sized species belonging to different clades may be differently vulnerable. Quaternary megafaunal extinctions in many parts of the world (e.g., North America, South America, Australia) affected animals larger than ~45 kilograms. Madagascar's "megafaunal" extinction eliminated all endemic species larger than 10 kilograms. This is a very low body mass extinction threshold, and it demands explanation. Perhaps the population sizes of Madagascar's "megafaunal" species prior to human arrival were low. Perhaps the human colonizers were unusually destructive. Or perhaps Madagascar's large-bodied species had characteristics that increased their vulnerability to extinction.

We know of no reason to believe that the behavior of the human colonizers was unusually destructive on Madagascar. Indeed, until fairly recently, human population density there was quite low. Furthermore, with the possible exception of ancestors of the Vezo of western Madagascar, the human colonizers were not hunter-gatherers. Could the particular vulnerability of Madagascar's mammals be related to certain unusual characteristics of these mammals themselves?

The Bet Hedgers

Madagascar's climate has been called hypervariable, and a number of researchers have suggested that the endemic fauna of Madagascar has long been shaped by that hypervariability (P. Wright 1999; Dewar and Richard 2007).[1] A reproductive strategy called "bet hedging" has evolved repeatedly and in very different taxa on Madagascar, and it is one that usually works well in unpredictable environments (Dewar and Richard 2007). Bet hedgers maximize adult survivorship at the expense of infant and juvenile mortality; they depend on high iteroparity (the production of offspring multiple times within the reproductive lifetime) to counteract the negative effects of bad years with exceptionally high infant and juvenile losses. For bet hedging to work as a reproductive "strategy," reproductive rates can be low and interbirth intervals long, but reproductive life spans must also be long and adult mortality must be relatively low. Bet hedging works well as long as adult mortality remains low, which is not the case when an invasive predator targeting fully grown individuals arrives. *Homo sapiens* is just such a predator.

Dewar and Richard (2007) argue that many of Madagascar's extant lemurs can be described as bet hedgers, and, citing prior demographic research by Richard et al. (2002), they showcase the Verreaux's sifaka (*Propithecus verreauxi*) as a quintessential example. Indeed, they maintain that a number of nonprimates endemic to Madagascar can be similarly described. The Malagasy jumping rat (*Hypogeomys antimena*), in sharp contrast to rodents in most other parts of the world, reproduces slowly. Its larger-bodied relative, *Hypogeomys australis,* suffered recent extinction, and was likely similar to *H. antimena* in its reproductive profile. Madagascar's endemic carnivores (belonging to the euplerid group) are also bet hedgers, giving birth once a year to a single offspring. The largest-bodied euplerid, *Cryptoprocta spelea,* also succumbed to extinction.

What about the giant extinct lemurs? Were they also bet hedgers? The only way to address this question is to estimate the reproductive rates of these species; actual reproductive rates can be measured only in extant species. Johnson (2002) attempted to do exactly this for the extinct members of nine mammalian families or superfamilies from various parts of the world, including the superfamily Lemuroidea (the lemurs). Considering only the extant members of each group separately, he regressed reproductive rate on body mass, and then estimated reproductive rates for extinct members of each group through extrapolation, on the basis of their reconstructed body masses. Of the nine mammalian groups that Johnson studied, the Lemuroidea had the lowest reproductive rates.

One can approach the problem in a different manner, employing the methods of dental microstructural analysis to estimate reproductive rates of extinct species. In recent years, dental microstructural analysis has been used to estimate life history parameters for extinct lemurs belonging to the three major clades, the Palaeopropithecidae, Archaeolemuridae, and Megaladapidae, and the methods are described in detail in Schwartz, Samonds, et al. 2002; Schwartz, Mahoney, et al. 2005; Godfrey, Schwartz, Samonds, et al. 2006; Godfrey, Schwartz, Sutherland, et al. 2008; and Catlett et al. 2010. Using these methods, one can measure accurately the age at first molar crown completion and rates of root growth, and, on the basis of these measures, one can estimate age at weaning, gestation length, minimum interbirth interval (IBI, equal to gestation length plus age at weaning), and maximum reproductive rate, which is merely the inverse of the minimum interbirth interval. This estimate of maximum reproductive rate assumes insignificant twinning or none at all, and thus does not pertain to species that regularly have multiple offspring per litter, as do many primates weighing less than 1 kilogram. If birthing is seasonal, then the minimum interbirth interval is equal to the smallest integer that accommodates the minimum IBI, and the maximum reproductive rate is the inverse of that integer.

The advantage of this method over Johnson's is that we can reconstruct reproductive rates of extinct species without reference to any presumed relationship between reproductive rate and body size; it also provides a ceiling for reproductive rates (they may have been lower in the extinct species, but they cannot have been higher, unless twinning was common). To estimate average (rather than maximum) reproductive rates, one can multiply maximum values by a correction factor, 0.72. This correction factor was derived by regressing observed reproductive rates (derived largely from Ross and Jones 1999) on the theoretical maxima for a sample of 36 primate species with adult female body masses of 1 kilogram or higher. The correlation between the two is high (0.84, P <0.001), and the slope of the relationship (0.72, P <0.001) provides the correction factor. The intercept was insignificantly different from 0, and was thus treated as 0.

Table 2 shows that Johnson's (2002) regression-based reproductive rates for the extinct lemurs, low as they are, are sometimes higher than the maxima derived through dental microstructural analysis. The discrepancy is greater if reproductive seasonality is assumed, or if the 0.72 correction factor is applied to the theoretical maxima. All of the extinct lemurs had minimum interbirth intervals in excess of one year, and if reproduction was seasonal, then females could not have given birth more often than once every two years. It is quite possible that reproduction was seasonal in extinct lemurs, as it is in the great majority of extant lemurs, including most of the larger-bodied species, as well as numerous large-bodied nonprimates (see, for example, Spady et al. 2007 on bears).

Thus it does appear that most (if not all) of the giant extinct lemurs were slow reproducers and likely bet hedgers. No matter how reproductive rates of the giant lemurs are reconstructed, they are lower than one offspring per year, sometimes markedly so. Exceptionally low reproductive rates would have made the giant lemurs vulnerable to the smallest increases in annual adult mortality at the hands of humans.

That these animals were bet hedgers makes sense in light of certain other characteristics of the lemur giants. Many (especially the megaladapids and palaeopropithecids) have been reconstructed as highly folivorous (i.e., consuming lots of foliage); their teeth have long, sharp crests for processing leaves (Jungers et al. 2002). They were also slow-moving (Jungers et al. 2002; Godfrey, Jungers, and Schwartz 2007; Walker et al. 2008) and may have been hypometabolic (able to survive on minimal energy input), as are living lemurs. Leaves are generally poor in nutrients, so many folivores have specialized guts, sleep a lot, and move slowly—all adaptations for living on minimal energy input. Excellent comparisons can be made to sloths, koalas, and giant pandas—some of the world's most lethargic mammals (see, for example, de Moura Filho et al. 1983; Kleiman 1983; Nagy and Martin 1985; Grand and Barboza 2001). As a group, specialized leaf eaters begin reproducing at an advanced age and may give birth every two to four years to small neonates that grow slowly and per-

Table 2. Johnson's (2002) regression-based reproductive rates for extinct lemurs from Madagascar's southwest, compared to those derived from dental microstructural analysis.*

Species, with estimated body mass (kg)	Gestation length (yrs)	Age at weaning (yrs)	IBI min. (yrs)	IBI min. seasonal (yrs)	Johnson's reprod. rate (offspring /year)	Max reprod. rate (offspring /year)	Max seasonal reprod. rate (offspring /year)	Corrected reprod. rate (offspring /year)
Archaeolemur majori (18.2)	0.48	1.67	2.15	3	0.73	0.47	0.33	0.34
Hadropithecus stenognathus (35.4)	0.52	3.04	3.56	4	0.63	0.28	0.25	0.20
Megaladapis edwardsi (85.1)	0.68	1.04	1.72	2	0.42	0.58	0.5	0.42
Mesopropithecus globiceps (11.3)	0.58	0.55	1.13	2	0.84	0.88	0.5	0.63
Palaeopropithecus ingens (41.5)	0.88	0.50	1.38	2	0.53	0.72	0.5	0.52

*For estimations of gestation length and age at weaning, see Catlett et al. 2010.
IBI min = gestation length + age at weaning
IBI min seasonal = next larger integer encompassing gestation length + age at weaning
    (= IBI minimum, assuming seasonality)
For calculation of Johnson's reproductive rate, see Johnson 2002 and supplementary data
Max reproductive rate = inverse of IBI minimum, with no assumption of birth seasonality
Max seasonal reproductive rate = inverse of IBI minimum, assuming birth seasonality
Corrected reproductive rate = (0.72)(max reproductive rate)

haps wean early, when they are still relatively small in body size. In a world in which adult survival depends on the ability to minimize energy expenditure, these adaptations can be very important, as gestation and especially lactation are energetically expensive.

Our data on dental chronology in the chimpanzee-sized *Palaeopropithecus ingens* show that this species experienced early molar crown completion, likely signifying early eruption of the permanent teeth coupled with early weaning (table 2). Prolonged prenatal molar crown formation suggests a long gestation period—longer than those of other giant lemurs and perhaps longer than that of *Homo sapiens* (Schwartz, Samonds, et al. 2002). Growth was relatively slow in *Palaeopropithecus* (Godfrey, Petto, et al. 2002); if weaning was early, weanlings were likely small in body size. This developmental profile (long gestation, short infancy, accelerated molar crown formation, slow growth) is also seen in *Propithecus* and other indriids; the decoupling of dental development and the growth of the body reflects a decoupling of age at ecological independence (weaning) and female age at first birth. The latter reflects overall growth rate. *Propithecus* has one of the slowest growth rates of all extant primates (slower than that of humans; see Mumby and Vinicius 2008); first births are also delayed in this species. This combination means that females can function independently as ecological adults long before they begin to reproduce. This certainly characterizes *Propithecus* and other indriids, and may have characterized *Palaeopropithecus* and other palaeopropithecids (Godfrey, Petto, et al. 2002; Godfrey and Jungers 2003; Schwartz, Samonds, et al. 2002).

Even more remarkable than *Palaeopropithecus* were the archaeolemurids, *Archaeolemur* and *Hadropithecus* (table 2). These are the extinct lemurs with the relatively largest brains (rivaling relative brain size of certain anthropoids of similar body size), and they contrast with the smaller-brained *Palaeopropithecus* (especially) and *Megaladapis* (to a lesser extent) in showing evidence of prolonged infancy and late weaning. The minimum estimated IBI derived from dental chronology work for *Archaeolemur* is 2.15 years (3, assuming reproductive seasonality), and the minimum estimated IBI for *Hadropithecus* is 3.56 years (4, assuming seasonality) (Godfrey, Semprebon, et al. 2005; Catlett et al. 2010). These values rival or exceed those of many large-bodied anthropoids. The archaeolemurids may have depended on certain foods that were difficult to access or to process, and their infants, like those of *Daubentonia*, may have needed extra time before weaning to acquire independent accessing or processing skills (Godfrey, Semprebon, et al. 2005) Clearly, adult archaeolemurids invested more energy in raising their young than did the palaeopropithecids or megaladapids, and prolonged time to weaning usually means delayed first reproduction. They may have been the most extreme in this regard.

Ongoing Extirpations and the Threat of More Extinctions

Approximately one-third of all recognized lemur species, including the vast majority of those species with data deemed sufficient for a valid field assessment, are considered threatened by the International Union for Conservation of Nature (IUCN). We know that the geographic ranges of many of these were larger in the recent past than they are now, sometimes considerably so

Table 3. Primates living today or in the recent past in the broad vicinity of the Beza Mahafaly Special Reserve (southwestern Madagascar), with measured or estimated body mass and conservation status according to the IUCN (http://www.iucnredlist.org).*

Species	Body Mass	Least concern	Threatened	Extinct
Microcebus griseorufus	60 g	x		
*Lepilemur petteri***	0.6 kg	Data deficient		
Lemur catta	2.2 kg		x	
Propithecus verreauxi	2.8 kg		x	
Mesopropithecus globiceps	11.3 kg			x
Pachylemur insignis	11.5 kg			x
Daubentonia robusta	14.2 kg			x
Archaeolemur majori	18.2 kg			x
Hadropithecus stenognathus	35.4 kg			x
Palaeopropithecus ingens	41.5 kg			x
Megaladapis madagascariensis	46.5 kg			x
Megaladapis edwardsi	85.1 kg			x

*For estimation of body mass of extinct lemurs, see Jungers, Demes, et al. 2008.
**The populations of Lepilemur in southern Madagascar (including that at the Beza Mahafaly Special Reserve) have long been called L. leucopus. On the basis of genetic evidence, Louis et al. 2006 distinguished the populations living in eastern and western portions of this range, calling the latter (at BMSR) L. petteri. More research on these populations is warranted to verify their distinctiveness and to assess their conservation status.

(Godfrey, Jungers, Simons, et al. 1999). Where forests or woodlands have disappeared entirely, so have lemurs. Many lemur species are restricted to wooded regions with low human populations, or to parcels of land accorded some kind of protected status by the government. In some places lemurs contribute significantly to the bushmeat market (e.g., Golden 2009), but even in places where hunting lemurs is culturally taboo (*fady*), they may be threatened.

We take as an example the region of the Beza Mahafaly Special Reserve (BMSR), where one of us (ER) has worked since 2003. A nearby subfossil site (Taolambiby) and others in the broader vicinity document the fauna in that area over the past several thousand years. Southwestern Madagascar appears to have been one of the last places to lose extinct lemurs and other megafauna. Although there is evidence of human presence in this region prior to two thousand years ago, human population expansion was relatively late here, as this is one of Madagascar's least hospitable habitats, due to its prolonged dry season. Within the Beza Mahafaly Special Reserve today, only four lemur species remain; of these, the two largest-bodied (and only diurnal) species, *Propithecus verreauxi* and *Lemur catta,* are currently considered vulnerable by the IUCN (IUCN 2009; see table 3); the other two primates (*Lepilemur petteri* and *Microcebus griseorufus*) are small-bodied and nocturnal. Table 3 presents the giant lemur species that are known to have lived in southwestern Madagascar in the recent past, some as recently as five hundred years ago. At Taolambiby, there is evidence of butchery of lemurs beginning 2,300 years ago and extending to historic times (Perez et al. 2005; B. Crowley, unpublished data).

Populations of the largest-bodied still-extant lemurs living at BMSR (sifakas and ringtailed lemurs) are in decline today. Lawler (2011) used genetic methods to pinpoint when that decline began for *Propithecus verreauxi*; his estimate was, remarkably, very close to the time of earliest evidence of butchery of giant lemurs (~2,300 years ago). Lawler found no evidence of a bottleneck or any abrupt change in population size, but rather a steady decline over a prolonged period. The implication is that the sifaka population in the southwest was much larger in the recent past than it is today. Crowley (unpublished) dated a number of cut-marked sifaka bones from Taolambiby; her data show that these animals were butchered over an extended period lasting at least a thousand years, and that this practice continued here until very recently—effectively, until the present (see also Burney, Burney, et al. 2004). Today, the local people of this region consider the consumption of sifaka meat fady. Clearly this wasn't always the case, and occasional killings of larger lemurs by people have been witnessed in recent years (Anne Axel, personal communication).

It is also clear that lemurs living in unprotected forests in the region of BMSR are not faring well. At Ihazoara, an unprotected forest only a few kilometers from the reserve and bordering a village bearing the same name, lemurs become increasingly scarce as one approaches the village. Indeed, only mouse lemurs survive in its immediate vicinity. Occasionally, a ringtailed lemur or sifaka will appear there (e.g., a lone *Lemur catta* from the reserve was sighted at the entrance to the village in 2003, and a sifaka carcass was observed in 2004 being consumed by village dogs).

The two forests that belong to BMSR proper (Parcel 1, gallery forest, and Parcel 2, spiny forest) differ in their degree of protection, Parcel 1 being fenced and generally monitored, and Parcel 2 being unfenced and rarely monitored. All four local lemur species live in both, but both the abundance and density of the two larger, diurnal species are greater in the forest with greater protection (Parcel 1), which is also nearer the Sakamena River, a tributary to the Onilahy River that is dry for eight months each year; see Axel and Maurer 2011.

The Least Endangered of Living Lemurs: Microcebus

Members of the genus *Microcebus,* the so-called mouse lemurs, are found all over Madagascar—wherever there is forest of any sort, and even in secondary habitats such as plantations. Today over a dozen species of *Microcebus* are recognized. Historic declines in population size and increased genetic differentiation have been documented for some of them; these can be attributed to recent forest loss and fragmentation (Olivieri et al. 2008). Some *Microcebus* species are likely endangered for these reasons (they are listed by the IUCN Red List as "data deficient"), but others are geographically widespread and among Madagascar's few lemur species considered unthreatened or of "least" conservation concern. This is the case for *M. griseorufus.*

At ~60 grams, *Microcebus griseorufus* is among the smallest-bodied of Madagascar's lemurs. Its reproductive profile stands in stark contrast to those of larger-bodied, "bet-hedging" lemurs; indeed, it may lie at the "fast" extreme of species within its own genus. Effectively, this species survives by being able to reproduce very rapidly; it maximizes reproduction at the expense of adult survivorship. Dewar and Richard (2007) maintain that hypervariable environments (such as are found in Madagascar) favor species at both reproductive extremes—i.e., those (like sifakas) that reproduce slowly but spread reproductive effort over a prolonged lifespan, and those (like mouse lemurs) that concentrate reproductive effort in a very short period of time. Mouse lemurs have two potential advantages over other lemurs. First, because they have very short generation times and multiple births per litter, they are better able to withstand high adult mortality rates. Second, because they are also very small in body size, they may not be hunted by humans at all.

Yet there is evidence that even *M. griseorufus* may be vulnerable to extinction in the not-too-distant future. This is despite the fact that *M. griseorufus* is practically ubiquitous in the spiny forests of the arid southwest, does not suffer there from human hunting, and actually prefers the widespread spiny and dry forests to rarer, wetter habitats in southern Madagascar (Yoder et al. 2002).

Table 4. Population turnover: Percentage of Microcebus individuals recaptured over extended periods*.

Species and sampling period	Forest	Known to be alive in at least two consecutive years	Known to be alive in at least three consecutive years	Known to be alive in at least four consecutive years
M. griseorufus 2003–2007	Ihazoara	5.1%	0	0
M. griseorufus 2003–2007	BMSR Parcel 1, gallery forest	14.6%	1.8%	1.2%
M. griseorufus 2003–2007	BMSR Parcel 2, spiny forest	13.7%	5.2%	1.3%
M. murinus 2000–2003	Mandena, southeast	13.0%	1.8%	0
M. rufus 2004–2008	Ranomafana, eastern rain forest	45.0%	25.2%	6.9%

*Sources: For percentages in the region of the Beza Mahafaly Special Reserve (and the unprotected forest, Ihazoara, near the reserve): Rasoazanabary 2011. For Mandena: Lahann et al., 2006. For Ranomafana: Marina Blanco, unpublished data.

Signs of trouble are threefold (Rasoazanabary 2011). First, mean monthly capture success rates for 23 months sampled over a period of five years (2003 to 2007) averaged 1.6 percent for the unprotected Ihazoara forest and 2.0 percent for forests within the reserve. If we take the period during which mouse lemurs are most likely to be captured (from July through October), the means are 2.1 percent for Ihazoara and 3.0 percent for the reserve forests. These rates are all very low, but they are lower, on average, in the unprotected than the protected forests.

Second, population turnover during the period from 2003 to 2007 was higher at Ihazoara than in the protected forests of BMSR. Only 5.1 percent of captured individuals were captured in two consecutive years at Ihazoara (cf. ~13 percent in the protected forests), and no individual was captured at Ihazoara over a period of three or more consecutive years (Table 4).

Third, there is evidence that *M. griseorufus* may be declining in its pre-ferred habitat, the spiny forest. Using capture-mark-recapture techniques, Rasoazanabary recorded a statistically significant drop in mouse lemur cap-tures in the spiny forest between 2003–2004 and 2006–2007 (*t*-test for mean monthly capture for the two periods excluding January through March, when mouse lemurs do not enter traps: $t=2.2$, $df=15$, $p=0.04$). Genetic data collected on individuals captured and released in 2003 (at the beginning of this study) suggested that the spiny forest housed the largest population of mouse lemurs, and the gallery forest the smallest (indeed, smaller in that year than Ihazo-ara) (Heckman et al. 2006). There was also evidence of bidirectional gene flow among all combinations of forest. By 2007, mouse lemur density had fallen measurably in the spiny forest and had risen in the better-protected gallery forest—a forest that does not present ideal habitat for this species of mouse lemur. The apparent reduction in mouse lemur population size in the spiny forest may be attributable to human disturbance. Despite its "protected" status, the spiny forest actually suffered more disturbance than did the gallery forest between 2003 and 2007 (Rasoazanabary 2004, 2011).

Human disturbance affects mouse lemurs both directly and indirectly. In the region of Beza Mahafaly, villagers cut fantsiolotra trees (*Alluaudia procera*) for house construction; planks can be used locally or sold at the market at Betioky (figure 8.2). Sasavy trees (*Salvadora angustifolia*) are sought because they attract bees, and people collect honey from bees' nests. To encourage the bees to leave their nests, people set the trees on fire. Unfortunately, both of these tree species are used by *M. griseorufus* for their own nests. Using nest checks of radio-collared individuals, ER found that spiny forest *Microcebus* use *Alluaudia procera* tree holes for their nests approximately 85 percent of the time. Because BMSR forests are "protected," local people are circumspect in entering them, often selecting times when the research camp director and sci-entific director are out of town, or mealtimes, when researchers have returned to the research camp. Loggers may arm themselves before entering the forest. The spiny forest is an easier target for loggers than the gallery forest, as it is rarely monitored. Occasionally in the spiny forest, large patches are cleared for maize agriculture. Cattle wander freely through the spiny forest, and oc-casionally the gallery forest, especially during the day, and sometimes are hid-den in the forests to protect them against local cattle thieves (*dahalo*). Hiding cattle in this manner (day and night) is common in unprotected forests, but in 2006, the fences were broken and the protected gallery forest was used for this purpose (Rasoazanabary 2011). Cattle hidden in the forest (especially at night) can disturb mouse lemur activities, driving them to change their nest

Figure 8.2. Planks cut from fantsiolotra trees (*Alluaudia procera*) for house construction in Beza Mahafaly region, Madagascar.
Photograph by Laurie R. Godfrey, ©2005.

sites, in turn making them more vulnerable to aerial predators and snakes. Owl predation in unprotected forests near BMSR was estimated by Goodman, Langrand, et al. (1993a, 1993b) to result in an annual mouse lemur population loss of ~25 percent.

The Biggest Challenge:
The Cultural Context of Conservation Policy

If it is generally the case that (1) lemurs are disappearing in zones of human/lemur interaction; (2) lemur populations are declining even where they are "protected"; and (3) small-bodied species that are in no sense bet hedgers are beginning to show signs of vulnerability, then the long-term prospects for lemur survival in the wild cannot be good. The act of according forests "protected" status, by itself, will not protect lemurs when local people are hostile, even weakly so, to conservation efforts.

Consider the Beza Mahafaly Project, in many ways a model for conservation policy, because ever since its inception in 1975 it has embraced active community involvement, multiple venues for improving local education, and minimal dependence on outside influence (Sussman and Ratsirarson 2006). Approximately one dozen local people are hired to work as guides and reserve staff, monitor the lemurs in the reserve, lead occasional tourists through the forest, and help visiting researchers. There is a sustained outreach effort. In 2003, reserve personnel began broadcasting biweekly FM radio programs from Betioky to convey the importance of BMSR forests and wildlife. Local villagers are regularly included in the broadcasts so that the conservation message is delivered using village views and voices. Reserve personnel also sponsor cultural festivals for neighboring villages to celebrate the environment and Malagasy culture. Finally, the staff invites local children and their teachers to the reserve each year for Earth Day, at which time the children visit the forests or go on field trips, draw pictures with conservation themes, and receive gifts. Principles of conservation are also taught in local schools. Three-quarters of the school-educated children have learned that forests should be protected to promote common, long-term conservation interests, rather than immediate personal gain (Rasoazanabary 2011). Fewer than one-third of those children who are not attending school spout the same ethic.

When asked their opinions of BMSR conservation policies, three-quarters of local village adults profess approval (Rasoazanabary 2011). They see foreigners as possible deliverers of goods or financial resources, funding for schooling for children, medical assistance, or employment for adults. Nevertheless, tensions exist between local villagers and reserve personnel, mainly because the local villagers view governmental claims to the land as fundamentally illegitimate (Richard and Dewar 2001; Primack and Ratsirarson 2005; Rasoazanabary 2011). The forests belong to the local people, being inherited directly from their ancestors. It matters not that the director, scientific director, and university students at BMSR are Malagasy. They are not from the local Mahafaly ethnic group. Reserve directors are considered foreigners with no right to the land. In fact, nonlocal Malagasy are less welcome than nonnationals. Newly arriving Malagasy researchers may be evaluated by local villagers for witchcraft, and they are more likely than nonnationals to suffer the wrath of local villagers should anything go wrong.

Many more children (43 percent) than adults (15 percent) have learned to read, and children who read are less likely to say they hunt wild animals within the boundaries of protected forests than those who don't. This does not imply that children do not hunt in the protected forest. Rather, a full two-thirds of the children who can't read profess to hunt in the protected forest, while fewer than half of those who can read make the same claim.

Hunting today does not generally target lemurs, as killing lemurs is *fady* in this region. However, lemurs have been killed at Beza Mahafaly just to make a point. Thus, for example, when Beza Mahafaly was first designated as a protected area, local villagers killed a number of sifakas and suspended their bodies in the middle of the road to register their anger at having been prohibited from using their traditional forests (Richard and Dewar 2001; Primack and Ratsirarson 2005).

Trees are sometimes felled for similar reasons. In 2003, some local villagers protested resource restrictions by cutting (but not removing) *Alluaudia* trees on a path in the spiny forest used by researchers. The damage was highly visible from the road, where it would be seen as a statement or warning. In 2005, some villagers who had been fined for cutting trees in protected forest reacted by cutting more trees to block the roads providing vehicle access to their village. They claimed the tree damage to be storm-related, although it was not.

Clearly, top-down conservation strategies are not solutions with long-term stability, and conservation education does not eradicate the sense of disempowerment that befalls villagers when forests they perceive as their property are declared "off limits" to local use. Of critical importance if the remaining lemurs are to be saved is that sound conservation policies be developed and championed by the local people.

Finally, one cannot pretend to address the cultural context of lemur endangerment and extinction without considering the effects of human policy decisions made far outside the boundaries of Madagascar. Global warming has become yet another factor threatening Madagascar's terrestrial biodiversity (Ingram and Dawson 2005; P. Wright 2007; Hannah et al. 2008; Raxworthy et al. 2008), and it is one over which the people of Madagascar have virtually no control. Global warming has not replaced other threats to species survival; it has merely compounded them. It will be impossible to predict future extinctions without studying the combined effects of these threats. Global warming increases both the frequency and the impact of extreme weather events, and thus increases environmental unpredictability. This makes the already-vulnerable species of Madagascar more vulnerable yet. Extreme weather events and changes in mean annual temperature and rainfall can also alter the needs and resources available to local people. Madagascar is currently identified as one of the world's hunger hotspots and is projected to remain so as global climates warm (Liu et al. 2008).

Summary and Conclusions

The island of Madagascar is an ideal subject for the study of extinction in "near time"—the effects of human invasion on endemic species. Because humans only arrived on this island a little more than two thousand years ago, the geological record of extinction events on Madagascar is excellent. Numerous "subfossil" sites allow for the documentation of the composition of wildlife before and after human arrival. A distanced perspective allows us to see ourselves as a species among others, and lets us evaluate the effects humans have had, as an invasive species, on endemic wildlife.

We have argued that humans have had a dramatic effect on Madagascar's wildlife—more extreme than at other places. As in other parts of the world, human-induced extinctions on Madagascar follow a particular pattern, with larger-bodied species disappearing first. However, the body mass extinction threshold is much lower on Madagascar than in other parts of the world. Taking lemurs as our primary example, we argue that the extreme effects witnessed here are due in part to the characteristics of Madagascar's endemic species themselves. Living on an island with an unpredictable climate, many of Madagascar's species had become bet hedgers. These are animals that maximize adult survivorship at the expense of infant and juvenile survival, and that depend on high iteroparity to counter episodes of exceptional infant and juvenile mortality. Bet hedging is normally advantageous in hypervariable environments. Unfortunately, bet hedging makes large-bodied species particularly vulnerable to extinction when annual rates of adult mortality climb, which is exactly what happens when an introduced predator, such as humans, targets large-bodied adult individuals. Small-bodied species have an advantage over large-bodied species by virtue of their shorter generation times.

Diverse lines of evidence suggest that Madagascar's giant extinct lemurs were bet hedgers par excellence. But even many of the smaller-bodied lemurs that are still alive can be characterized as bet hedgers. Like-sized species can differ markedly in their interbirth intervals, reproductive rates, and generation times. When compared to nonprimates outside Madagascar, lemurs have exceptionally low reproductive rates for their body sizes. This may explain why the body mass threshold for "megafaunal" extinction is so low on Madagascar, and why so many of the remaining smaller-bodied lemurs are threatened.

Humans can affect annual mortality rates directly (via hunting) or indirectly (via forest fragmentation, selective use of forest resources, and other general uses of the forests). All of these are implicated as factors synergistically contributing to species' range reductions and extinctions over the past two millennia. The process has been slow but inexorable. There is some indication that hunting was relatively more important in the past, and that deforestation and forest fragmentation are more important today. There is also no indication that human-mediated extinctions have ended on Madagascar; indeed, they may be accelerating. Increasingly, smaller-bodied species, including some that cannot be considered bet hedgers at all, are affected, and lemurs are threatened even in places where hunting them is culturally forbidden. If it is generally the case that lemurs are disappearing in zones of human/lemur interaction, and if living lemur populations are declining even where they are "protected," then the long-term prospects for lemur survival in the wild are poor.

Finally, we argue that conservation work today faces enormous obstacles; top-down "fortress conservation" ideology still dominates on the Great Red Island, but it has come under the critical scrutiny of political ecologists and cultural anthropologists (Brockington 2002; Kaufmann 2006; J. Harper 2008; Ferguson 2009). Conservationists themselves appreciate its failures and increasingly advocate at least a partial shift from top-down to bottom-up conservation policies. Whereas it is not clear that bottom-up conservation policies will work, it is clear that no conservation policy can succeed in the long term without the support of the local people. At the very least, some sort of combined bottom-up, top-down policy building is imperative.

NOTES

1. Dewar and Richard (2007) have shown that the climate of Madagascar is indeed hypervariable. Using Colwell's (1974) criteria of rainfall constancy (C—the extent to which rainfall is constant over the year from month to month) and contingency (M—the extent to which rainfall during any selected month varies from year to year) to describe predictability, they compared 15 dry and wet forests on Madagascar to 15 dry and wet forests in continental Africa, matching annual rainfall. They found that Madagascar's eastern rainforests have high constancy but low contingency, while sites in the center, north, south, and west have high contingency but low constancy (due to their prolonged dry seasons). By one criterion or the other, Madagascar's environments really are different from those of continental Africa, which exhibit moderate constancy and moderate contingency, and thus cluster in the middle on a bivariate plot.

REFERENCES

Agarwal, Deepak K., John A. Silander, Jr., Alan E. Gelfand, Robert E. Dewar, and John G. Mickelson, Jr.
2005 Tropical Deforestation in Madagascar: Analysis Using Hierarchical, Spatially Explicit, Bayesian Regression Models. Ecological Modelling 185:105–131.

Ali, Jason R., and Matthew Huber
2010 Mammalian Biodiversity on Madagascar Controlled by Ocean Currents. Nature 463:653–656.

Allibert, Claude
2008 Austronesian Migration and the Establishment of the Malagasy Civilization: Contrasted Readings in Linguistics, Archaeology, Genetics and Cultural Anthropology. Diogenes 218:7–16.

Axel, Anne C., and Brian A. Maurer
2011 Lemurs in a Complex Landscape: Mapping Species Density in Subtropical Dry Forests of Southwestern Madagascar Using Data at Multiple Levels. American Journal of Primatology 7(1):38–52.

Blench, Roger
2007 New Palaeozoogeographical Evidence for the Settlement of Madagascar. Azania 42:69–82.

Bond, William J., John A. Silander, Jr., Jeannine Ranaivonasy, and Joelisoa Ratsirarson
2008 The Antiquity of Madagascar's Grasslands and the Rise of C_4 Grassy Biomes. Journal of Biogeography 35:1743–1758.

Brochu, Christopher A.
2007 Morphology, Relationships, and Biogeographical Significance of an Extinct Horned Crocodile (Crocodylia, Crocodylidae) from the Quaternary of Madagascar. Zoological Journal of the Linnean Society 150(4):835–863.

Brockington, Dan
2002 Fortress Conservation: The Preservation of the Mkomazi Game Reserve, Tanzania. Bloomington: Indiana University Press.

Brockman, Diane K., Laurie R. Godfrey, Luke J. Dollar, and Joelisoa Ratsirarson
2008 Evidence of Invasive *Felis silvestris* Predation on *Propithecus verreauxi* at Beza Mahafaly Special Reserve, Madagascar. International Journal of Primatology 29:135–152.

Brook, Barry W., and David M. J. S. Bowman
2005 One Equation Fits Overkill: Why Allometry Underpins both Pre-historic and Modern Body Size-Biased Extinctions. Population Ecology 47:137–141.

Burney, David A.
1993 Late Holocene Environmental Changes in Arid Southwestern Madagascar. Quaternary Research 28:274–280.

Burney, David A., Lida P. Burney, Laurie R. Godfrey, William L. Jungers, Steven M. Goodman, Henry T. Wright, and A. J. Timothy Jull
2004 A Chronology for Late Prehistoric Madagascar. Journal of Human Evolution 47(1–2):25–63.

Burney, David A., Helen F. James, Fred V. Grady, J.-G. Rafamantanantsoa, Ramilisonina, Henry T. Wright, and James B. Cowart
1997 Environmental Change, Extinction, and Human Activity: Evidence from Caves in NW Madagascar. Journal of Biogeography 24:755–767.

Burney, David A., and Ramilisonina
1998 The Kilopilopitsofy, Kidoky, and Bokyboky: Accounts of Strange Animals from Belo-Sur-Mer, Madagascar, and the Megafaunal "Extinction Window." American Anthropologist 100(4):957–966.

Burney, David A., Guy S. Robinson, and Lida P. Burney
2003 Sporormiella and the Late Holocene Extinctions in Madagascar. Proceedings of the National Academy of Sciences 100(19):10800–10805.

Catlett, Kierstin K., Gary T. Schwartz, Laurie R. Godfrey, and William L. Jungers
2010 "Life History Space": A Multivariate Analysis of Life History Variation in Extant and Extinct Malagasy Lemurs. American Journal of Physical Anthropology 142:391–404.

Colwell, Robert K.
1974 Predictability, Constancy and Contingency of Periodic Phenomena. Ecology 55:1148–1153.

Crowley, Brooke E.
2010 A Refined Chronology of Prehistoric Madagascar and the Demise of the Megafauna. Quaternary Science Reviews 29:2591–2603.

de Moura Filho, Albero Galvão, Sara Espe Huggins, and Salustiano Gomes Lines
1983 Sleep and Waking in the Three-Toed Sloth, Bradypus tridactylus. Comparative Biochemistry and Physiology, part A, 76(2):345–355.

Dewar, Robert E.
  2003 Relationship between Human Ecological Pressure and the Vertebrate
  Extinctions. *In* The Natural History of Madagascar. Steven M. Goodman
  and Jonathan P. Benstead, eds. Pp. 119–122. Chicago: University of Chicago
  Press.

Dewar, Robert E., and Alison F. Richard
  2007 Evolution in the Hypervariable Environment of Madagascar. Proceedings of the National Academy of Sciences 104(34):13723–13727.

Dewar, Robert E., and Wright, Henry T.
  1993 The Culture History of Madagascar. Journal of World Prehistory
  7(4):417–466.

Faure, M., and Céline Guérin
  1990 *Hippopotamus laloumena* nov. sp. The 3rd Holocene *Hippopotamus*
  Species of Madagascar. Comptes Rendus de l'Académie des Sciences, série 2,
  310(9):1299–1305.

Ferguson, Barry
  2009 REDD Comes into Fashion in Madagascar. Madagascar Conservation
  & Development 4(2):132–137.

Flacourt, Etienne de
  1658 Histoire de la Grand Isle Madagascar avec une Relation de Ce Qui
  S'est Passé en Années 1655, 1656, and 1657 Non Encore Vue par la Première
  Impression. Paris: Chez Pierre Bien-Fait.

Gasse, Françoise, and Elise Van Camp
  1998 A 40,000–Yr Pollen and Diatom Record from Lake Tritrivakely, Madagascar, in the Southern Tropics. Quaternary Research 49(3):299–311.
  2001 Late Quaternary Environmental Changes from a Pollen and Diatom
  Record in the Southern Tropics (Lake Tritrivakely, Madagascar). Palaeo-
  geography, Palaeoclimatology, Palaeoecology 167:287–308.

Godfrey, Laurie R.
  1986 The Tale of the Tsy-aomby-aomby. The Sciences 1986:49–51.

Godfrey, Laurie R., and Mitchell T. Irwin
  2007 The Evolution of Extinction Risk: Past and Present Anthropogenic
  Impacts on the Primate Communities of Madagascar. Folia Primatologica
  78(5–6):405–419.

Godfrey, Laurie R., and William L. Jungers
  2003 The Extinct Sloth Lemurs of Madagascar. Evolutionary Anthropology
  12(6):252–263.

Godfrey, Laurie R., William L. Jungers, Kaye E. Reed, Elwyn L. Simons, and Prithijit S. Chatrath
1997 Subfossil Lemurs: Inferences about Past and Present Primate Communities. *In* Natural Change and Human Impact in Madagascar. Steven M. Goodman and Brian Patterson, eds. Pp. 218–256. Washington, D.C.: Smithsonian Institution Press.

Godfrey, Laurie R., William L. Jungers, and Gary T. Schwartz
2007 Ecology and Extinction of Madagascar's Subfossil Lemurs. *In* Lemurs: Ecology and Adaptation. Lisa Gould and Michelle Sauther, eds. Pp. 41–64. New York: Springer.

Godfrey, Laurie R., William L. Jungers, Elwyn L. Simons, Prithijit S. Chatrath, and Berthe Rakotosamimanana
1999 Past and Present Distributions of Lemurs in Madagascar. *In* New Directions in Lemur Studies. Berthe Rakotosamimanana, Hanta Rasamimanana, Jörg U. Ganzhorn, and Steven M. Goodman, eds. Pp. 19–53. New York: Plenum.

Godfrey, Laurie R., Andrew J. Petto, and Michael R. Sutherland
2002 Dental Ontogeny and Life-History Strategies: The Case of the Giant Extinct Indrioids of Madagascar. *In* Reconstructing Behavior in the Primate Fossil Record. J. Michael Plavcan, Richard F. Kay, William L. Jungers, and Carl P. Van Schaik, eds. Pp. 113–157. New York: Kluwer Academic/Plenum.

Godfrey, Laurie R., Karen E. Samonds, William L. Jungers, Michael R. Sutherland, and Mitchell T. Irwin
2004 Ontogenetic Correlates of Diet in Malagasy Lemurs. American Journal of Physical Anthropology 123:250–276.

Godfrey, Laurie R., Gary T. Schwartz, Karen E. Samonds, William L. Jungers, and Kierstin K. Catlett
2006 The Secrets of Lemur Teeth. Evolutionary Anthropology 15(4):142–154.

Godfrey, Laurie R., Gary T. Schwartz, Michael R. Sutherland, William L. Jungers, and Mitchell T. Irwin
2008 Life History and Body Size as Determinants of Quaternary Mammalian Species Loss. Paper presented at the 22nd Congress of the International Primatological Society, Edinburgh, August 6. Abstract no. 425.

Godfrey, Laurie R., Gina M. Semprebon, Gary T. Schwartz, David A. Burney, William L. Jungers, Erin Flanagan., Frank P. Cuozzo, and Stephen J. King
2005 New Insights into Old Lemurs: The Trophic Adaptations of the Archaeolemuridae. International Journal of Primatology 26:825–854.

Golden, Christopher D.
2009 Bushmeat Hunting and Use in the Makira Forest, North-Eastern Madagascar: A Conservation and Livelihoods Issue. Oryx 43(3):386–392.

Goodman, Steven M.
1994 Description of a New Species of Subfossil Eagle from Madagascar, *Stephanoaetus* (Aves, Falconiformes), from the Deposits of Ampasambazimba. Proceedings of the Biological Society of Washington 107(3):421–428.

Goodman, Steven M., Olivier Langrand, and Christopher J. Raxworthy
1993a Food Habits of the Madagascar Long-Eared Owl *Asio madagascariensis* in Two Habitats in Southern Madagascar. Ostrich 64:79–85.
1993b The Food Habits of the Barn Owl *Tyto alba* at Three Sites on Madagascar. Ostrich 64:160–171.

Goodman, Steven M., and Daniel Rakotondravony
1996 The Holocene Distribution of *Hypogeomys* (Rodentia: Muridae: Nesomyinae) on Madagascar. *In* Biogéographie de Madagascar. Wilson R. Lourenço, ed. Pp. 283–293. Paris: ORSTOM Éditions.

Goodman, Steven M., and Lucien M. A. Rakotozafy
1995 Evidence for the Existence of Two Species of *Aquila* on Madagascar during the Quaternary. Geobios 28(2):241–246.

Goodman, Steven M., Rodin M. Rasoloarison, and Jörg U. Ganzhorn
2004 On the Specific Identification of Subfossil *Cryptoprocta* (Mammalia, Carnivora) from Madagascar. Zoosystema 28(1):129–143.

Goodman, Steven M., Natalie Vasey, and David A. Burney
2007 Description of a New Species of Subfossil Shrew Tenrec (Afrosoricida: Tenrecidae: *Microgale*) from Cave Deposits in Southeastern Madagascar. Proceedings of the Biological Society of Washington 120(4):367–376.

Grand, Theodore I., and Perry S. Barboza
2001 Anatomy and Development of the Koala, *Phascolarctos cinereus:* An Evolutionary Perspective on the Superfamily Vombatoidea. Anatomy and Embryology 203(3):211–223.

Grandidier, Guillaume
1902 Une Mission dans la Région Australe de Madagascar en 1901. La Géographie 6:1–16.

Hannah, Lee, Radhika Dave, Porter P. Lowry II, Sandy Andelman, Michele Andrianarisata, Luciano Andriamaro, Alison Cameron, Robert Himmans, Claire Kremen, James MacKinnon, Harison Hanitriniaina Randrianasolo, Sylvie Andriambololonera, Andriamandimbisoa Razafimpahanana, Herilala Randriamahazo, Jeannicq Randrianarisoa, Philippe Razafinjatovo, Chris Raxworthy, George E. Schatz, Mark Tadross, and Lucienne Wilmé
    2008 Opinion Piece. Climate Change Adaptation in Madagascar. Biology Letters 4(5):590–594.

Harper, Grady J., Marc K. Steininger, Compton J. Tucker, Daniel Juhn, and Frank Hawkins
    2007 Fifty Years of Deforestation and Forest Fragmentation in Madagascar. Environmental Conservation 34(4):325–333.

Harper, Janice
    2008 The Environment of Environmentalism: Turning the Ethnographic Lens on a Conservation Project. In Greening the Great Red Island: Madagascar in Nature and Culture. Jeffrey C. Kaufmann, ed. Pp. 241–274. Pretoria: Africa Institute of South Africa.

Hébert, J-C.
    1998 La Relation du Voyage à Madagascar de Ruelle (1665–1668). Études Ocean Indien 25–26:9–94.

Heckman, Kellie L., Emilienne Rasoazanabary, Erica Machlin, Laurie R. Godfrey, and Anne D. Yoder
    2006 Incongruence between Genetic and Morphological Diversity in Microcebus griseorufus of Beza Mahafaly. BMC Evolutionary Biology 6, article no. 98.

Hurles, Matthew E., Bryan C. Sykes, Mark A. Jobling, and Peter Forster
    2005 The Dual Origin of the Malagasy in Island Southeast Asia and East Africa: Evidence from Maternal and Paternal Lineages. American Journal of Human Genetics 76(5):894–901.

Ingram, Jane C., and Terence P. Dawson
    2005 Climate Change Impacts and Vegetation Response on the Island of Madagascar. Philosophical Transactions of the Royal Society, part A, 363:55–59.

IUCN (International Union for Conservation of Nature)
    2009 IUCN Red List of Threatened Species. Version 2009.2. http://www.iucnredlist.org/, accessed December 22, 2009.

Johnson, Christopher N.
    2002 Determinants of Loss of Mammal Species during the Late Quaternary "Megafauna" Extinctions: Life History and Ecology, but Not Body Size. Proceedings of the Royal Society of London, series B, Biological Sciences 269(1506):2221–2227.

Jungers, William L., Laurie R. Godfrey, Elwyn L. Simons, Roshna E. Wunderlich, Brian G. Richmond, and Prithijit S. Chatrath
  2002 Ecomorphology and Behavior of Giant Extinct Lemurs from Madagascar. *In* Reconstructing Behavior in the Primate Fossil Record. J. Michael Plavcan, Richard F. Kay, William L. Jungers, and Carl P. Van Schaik, eds. Pp. 371–411. New York: Kluwer Academic/Plenum.

Jungers, William L., Brigitte Demes, and Laurie R. Godfrey
  2008 How Big Were the "Giant" Extinct Lemurs of Madagascar? *In* Elwyn Simons: A Search for Origins. John G. Fleagle and Christopher C. Gilbert, eds. Pp. 343–360. New York: Springer.

Kaufmann, Jeffrey C.
  2001 La Question des Raketa: Colonial Struggles with Prickly Pear Cactus in Southern Madagascar, 1900–1923. Ethnohistory 48(1–2):89–124.
  2006 The Sad Opaqueness of the Environmental Crisis in Madagascar. Conservation and Society 4(2):179–183.

Kleiman, Devra G.
  1983 Ethology and Reproduction of Captive Giant Pandas (*Ailuropoda melanoleuca*). Zeitschrift für Tierpsychologie 62(1):1–46.

Koch, Paul L., and Anthony D. Barnosky
  2006 Late Quaternary Extinctions: State of the Debate. Annual Review of Ecology, Evolution, and Systematics 37:215–250.

Lahann, Petra, Jutta Schmid, and Jörg U. Ganzhorn
  2006 Geographic Variation in Life History Traits of *Microcebus murinus* in Madagascar: Resource Seasonality or Bergmann's Rule? International Journal of Primatology 27(4):983–999.

Lawler, Richard R.
  2011 Historical Demography of a Wild Lemur Population (*Propithecus verreauxi*) in Southwest Madagascar. Population Ecology 53(1):229–240.

Liu, Junguo, Steffen Fritz, C. F. A. Van Wesenbeeck, Michael Fuchs, Liangzhi You, Michael Obersteiner, and Hong Yang
  2008 A Spatially Explicit Assessment of Current and Future Hotspots of Hunger in Sub-Saharan Africa in the Context of Global Change. Global and Planetary Change 64:222–235.

Louis, Edward E., Jr., Shannon E. Engberg, Runhua Lei, Huimin Geng, Julie A. Sommer, Richard Randriamampionona, Jean C. Randriamanana, John R. Zaonarivelo, R. Andriantompohavana, Gisèle Randria, P. Prosper, B. Ramaromilanto, G. Rakotoarisoa, A. Rooney, and R. A. Brenneman.
    2006 Molecular and Morphological Analyses of the Sportive Lemurs (Family Megaladapidae: Genus *Lepilemur*) Reveals 11 Previously Unrecognized Species. Special Publications of the Museum of Texas Tech University 49:1–49.

Lyons, S. Kathleen, Felisa A. Smith, and James H. Brown
    2004 Of Mice, Mastodons, and Men: Human-Mediated Extinctions on Four Continents. Evolutionary Ecology Research 6:339–358.

MacPhee, Ross D. E.
    1994 Morphology, Adaptations and Relationships of *Plesiorycteropus,* and a Diagnosis of a New Order of Eutherian Mammals. Bulletin of the American Museum of Natural History 220:1–214.

MacPhee, Ross D. E., ed.
    1999 Extinctions in Near Time: Causes, Contexts, and Consequences. New York: Springer.

Magnavita, Carlos
    2006 Ancient Humped Cattle in Africa: A View from the Chad Basin. African Archaeological Review 23:55–84.

Mein, Pierre, Frank Sénégas, Dominique Gommery, Beby Ramanivosoa, Hervé Randrianantenaina, and Patrice Kerloc'h
    2010 Nouvelles Espèces Subfossiles de Rongeurs du Nord-Ouest de Madagascar. Comptes Rendus Palevol 9:101–112.

Muldoon, Kathleen M., Donald D. de Blieux, Elwyn L. Simons, and Prithijit S. Chatrath.
    2009 The Subfossil Occurrence and Paleoecological Significance of Small Mammals at Ankilitelo Cave, Southwestern Madagascar. Journal of Mammalogy 90(5):1111–1131.

Muldoon, Kathleen M., and Elwyn L. Simons
    2007 Ecogeographic Size Variation in Small-Bodied Subfossil Primates from Ankilitelo, Southwestern Madagascar. American Journal of Physical Anthropology 134:152–161.

Mumby, Hannah, and Lucio Vinicius
    2008 Primate Growth in the Slow Lane: A Study of Inter-species Variation in the Growth Constant A. Evolutionary Biology 35(4):287–295.

Nagy, Kenneth A., and Robert W. Martin
1985 Field Metabolic Rate, Water Flux, Food Consumption and Time Budget of Koalas, *Phascolarctos cinereus* (Marsupialia, Phascolarctidae) in Victoria. Australian Journal of Zoology 33(5):655–665.

Olivieri, Gillian L., Vitor Sousa, Lounès Chikhi, and Ute Radespiel
2008 From Genetic Diversity and Structure to Conservation: Genetic Signature of Recent Population Declines in Three Mouse Lemur Species (*Microcebus* spp.). Biological Conservation 141(5):1257–1271.

Perez, Ventura R., Laurie R. Godfrey, Malgosia Nowak-Kemp, David A. Burney, Jonah Ratsimbazafy, and Natalie Vasey
2005 Evidence of Early Butchery of Giant Lemurs in Madagascar. Journal of Human Evolution 49(6):722–742.

Population Reference Bureau
2010 World Population Data Sheet. Washington, D.C.: Population Reference Bureau. http://www.prb.org/pdf10/10wpds_eng.pdf, accessed April 20, 2011.

Primack, Richard B., and Joelisoa Ratsirarson
2005 Principe de Base de la Conservation de la Biodiversité. Antananarivo, Madagascar: MacArthur and Université d'Antananarivo, École Supérieure des Sciences Agronomiques.

Rasoazanabary, Emilienne
2004 A Preliminary Study of Mouse Lemurs in the Beza Mahafaly Special Reserve, Southwest Madagascar. Lemur News 9:4–7.
2011 The Human Factor in Mouse Lemur Conservation: Local Resource Utilization and Habitat Disturbance at Beza Mahafaly Special Reserve, SW Madagascar. Ph.D. dissertation, University of Massachusetts, Amherst.

Raxworthy, Christopher J., Richard G. Pearson, Nirhy Rabibisoa, Andry M. Rakotondrazafy, Jean-Baptiste Ramanamanjato, Achille P. Raselimanana, Shenghai Wu, Ronald A. Nussbaum, and Dáithí A. Stone
2008 Extinction Vulnerability of Tropical Montane Endemism from Warming and Upslope Displacement: A Preliminary Appraisal for the Highest Massif in Madagascar. Global Change Biology 14:1703–1720.

Richard, Alison F., and Robert E. Dewar
2001 Politics, Negotiation and Conservation: A View from Madagascar. *In* African Rain Forest Ecology and Conservation: An Interdisciplinary Perspective. William Weber, Lee J. T. White, Amy Vedder, and Lisa Naughton-Treves, eds. Pp. 535–546. New Haven: Yale University Press.

Richard, Alison F., Robert E. Dewar, Marion Schwartz, and Joelisoa Ratsirarson
2002 Life in the Slow Lane? Demography and Life History of Male and Female Sifaka (*Propithecus verreauxi verreauxi*). Journal of Zoology, London 256:421–436.

Ross, Carolyn, and Kate E. Jones
1999 Socioecology and the Evolution of Primate Reproduction. *In* Comparative Primate Socioecology. Phyllis C. Lee, ed. Pp. 73–110. Cambridge: Cambridge University Press.

Samonds, Karen E.
2007 Late Pleistocene Bat Fossils from Anjohibe Cave, Northwestern Madagascar. Acta Chiropterologica 9(1):39–65.

Schwartz, Gary T., Patrick Mahoney, Laurie R. Godfrey, Frank P. Cuozzo, William L. Jungers, and Gisèle F. N. Randria
2005 Dental Development in *Megaladapis edwardsi* (Primates, Lemuriformes): Implications for Understanding Life History Variation in Subfossil Lemurs. Journal of Human Evolution 49(6):702–742.

Schwartz, Gary T., Karen E. Samonds, Laurie R. Godfrey, William L. Jungers, and Elwyn L. Simons
2002 Dental Microstructure and Life History in Subfossil Malagasy Lemurs. Proceedings of the National Academy of Sciences 99:6124–6129.

Smith, Andrew P., N. Horning, and D. Moore
1997 Regional Biodiversity Planning and Lemur Conservation with GIS in Western Madagascar. Conservation Biology 11(2):498–512.

Sodhi, Navjot S.
2008 Tropical Biodiversity Loss and People—A Brief Review. Basic and Applied Ecology 9:93–99.

Spady, Thomas J., Donald G. Lindburg, and Barbara S. Durrant
2007 Evolution of Reproductive Seasonality in Bears. Mammal Review 37(1):21–53.

Sussman, Robert W., and Joelisoa Ratsirarson
2006 Beza Mahafaly Special Reserve: A Research Site in Southwestern Madagascar. *In* Ringtailed Lemur Biology: *Lemur catta* in Madagascar. A. Jolly, R. W. Sussman, N. Koyama, and H. Rasamimanana, eds. Pp. 41–49. New York: Springer.

Tofanelli, Sergio, Stefania Bertoncini, Loredana Castrì, Donata Luiselli, Francesc Calafell, Giuseppi Donati, and Giorgio Paoli
    2009 On the Origins and Admixture of Malagasy: New Evidence from High Resolution Analyses of Paternal and Maternal Lineages. Molecular Biology and Evolution 26(9):2109–2124.

Virah-Sawmy, Malika, Katherine J. Willis, and Lindsey Gillson
    2009 Threshold Response of Madagascar's Littoral Forest to Sea-Level Rise. Global Ecology and Biogeography 18(1):98–110.
    2010 Evidence for Drought and Forest Declines during the Recent Megafaunal Extinctions in Madagascar. Journal of Biogeography 37:506–519.

Walker, Alan, Timothy M. Ryan, Mary T. Silcox, Elwyn L. Simons, and Fred Spoor
    2008 The Semicircular Canal System and Locomotion: The Case of Extinct Lemuroids and Lorisoids. Evolutionary Anthropology 17:135–145.

Wells, Neil A.
    2003 Some Hypotheses on the Mesozoic and Cenozoic Paleoenvironmental History of Madagascar. In The Natural History of Madagascar. S. M. Goodman and J. P. Benstead, eds. Pp. 16–34. Chicago: University of Chicago Press.

Wright, Henry T., and Jean-Aimé Rakotoarisoa
    2003 The Rise of Malagasy Societies: New Developments in the Archaeology of Madagascar. In The Natural History of Madagascar. Steven M. Goodman and Jonathan P. Benstead, eds. Pp. 112–119. Chicago: University of Chicago Press.

Wright, Patricia C.
    1999 Lemur Traits and Madagascar Ecology: Coping with an Island Environment. Yearbook of Physical Anthropology 42:31–72.
    2007 Considering Climate Change Effects in Lemur Ecology and Conservation. In Lemurs, Ecology and Adaptation. Lisa Gould and Michelle L. Sauther, eds. Pp. 385–401. New York: Springer.

Yoder, Anne D., Melissa M. Burns, and Fabien Génin
    2002 Molecular Evidence of Reproductive Isolation in Sympatric Sibling Species of Mouse Lemurs. International Journal of Primatology 23:1335–1343.

Youssouf Jacky, Ibrahim, and Emilienne Rasoazanabary
    2008 Discovery of *Macrotarsomys bastardi* at Beza Mahafaly Special Reserve, Southwest Madagascar, with Observations on the Dynamics of Small Mammal Interactions. Madagascar Conservation & Development 3(1):31–37.

9. DISAPPEARING WILDMEN

CAPTURE, EXTIRPATION, AND EXTINCTION AS REGULAR COMPONENTS OF REPRESENTATIONS OF PUTATIVE HAIRY HOMINOIDS

Gregory Forth

The evolutionary path is almost by definition littered with extinctions. New species arise as others cease to exist, either entirely or by changing sufficiently to count as new species. This is as true of the evolution of *Homo* and other hominid genera as to any other animal kinds. This chapter concerns the figure of the "wildman," a name employed, somewhat arbitrarily, to refer to images of coarse-featured, usually hairy hominoids living a crude existence in desolate places far from human habitations. Usually considered a reference to an imaginary being, the term "wildman" comes from a European literary and artistic tradition, one mostly known from late mediaeval culture but deriving from pre-Christian folklore and surviving in folk traditions to the twentieth century. However, it cannot be overstressed that similar images of hairy subhumans are to be found in many parts of the world, and occur independently of the European tradition in the cultures and oral history of a number of small-scale non-Western societies in Asia, Africa, Oceania, and elsewhere.

In the last few years, one such non-Western image has received some attention from Western media. This is the figure called *ebu gogo*, known to inhabitants of the central part of Flores Island in eastern Indonesia, and specifically to the ethnolinguistic group named Nage. Described as shorter than local humans (standing 1–1.5 meters, according to most estimates), these ebu gogo have moreover been linked with the recently discovered subfossil hominin designated *Homo floresiensis* (and rather less fortunately dubbed "the hobbit"). Both ebu gogo and *Homo floresiensis* are considered to be extinct, and the ebu gogo are claimed to have been rendered extinct by local humans.

Coincidentally, it was the ebu gogo of Flores who catalyzed my interest in wildman images. I first encountered the wildman as an ethnographic problem during fieldwork I conducted on the eastern Indonesian islands of Sumba and Flores during the 1970s and 1980s. On both islands, local people described categories of hairy hominoids known to their relatively recent ancestors and, in some instances, possibly surviving to the present day. Although significantly different from better-known creatures supposedly encountered in other parts of the world, the Indonesian images immediately put me in mind of such figures as the North American sasquatch or "bigfoot" and the Himalayan yeti. Years later, I had the opportunity to initiate comparative research into categories of hominoids recognized by various cultures, using Southeast Asian exemplars as my focus. The main outcome has been a recent book (Forth 2008). One incentive for pursuing the topic was the fact that such figures have largely been ignored by anthropologists. Wildmen have received somewhat more attention from historians and folklorists, who have mostly been concerned with the European variety while making occasional comparative reference to yeti and sasquatch. But, like anthropologists, they typically assume that the categories refer to entirely imaginary beings, as indeed do most natural scientists, the large majority of whom do not consider the reputed existence of such creatures to merit any sort of empirical investigation.

A very different approach characterizes cryptozoology, literally the "science of hidden animals." Cryptozoologists—many of whom are trained zoologists with other research agendas besides cryptozoology—start from the position that reports of creatures which appear not to fit known species, rather than being the products of fantasy, may reflect species that have yet to be "discovered," documented, and thus recognized by modern science. However, cryptozoology too often displays a Western bias, and also what might be called a "visual" bias, relying largely on the accounts of putative observations by Europeans or Euro-Americans. Accordingly, cryptozoologists have paid little attention to categories and representations of non-Western peoples, and, being mostly focused on what are reported as contemporary or recent sightings of "crypto-species," they have insufficiently considered how the categories figure in the oral traditions and folk zoological knowledge of local people, especially as these could inform interaction and relations between humans and some empirical creatures, including creatures that have since disappeared from a given region. Of course, in one sense, they can hardly be blamed for this, in view of the scant attention given to such categories by ethnographers and cultural anthropologists.

The fact that local people describe them as either extinct or much rarer than in previous times is a particular point of interest in the study of unattested hominoids, and none more so than Southeast Asian categories like the ebu gogo of Flores. Not only does the theme of extinction or near extinction obviously contribute to the naturalistic (or nonspiritual) quality of this sort of representation, but, as comparative research has revealed, themes of hunting, capture, and indeed killing and extirpation of wildmen by humans are common wherever wildman categories are found. One question that arises, then, is to what extent hominoidal images—often articulated in current reports or rumors of encounters with unfamiliar creatures as well as in folktales and legends—might not be pure fantasies, perhaps serving some social or ideological function, but might rather reflect actual encounters with some sort of natural beings either human or nonhuman: encounters that result in a significant reduction of the latter, if not their complete destruction.

This is not to claim that all wildman images attest to human encounters with scientifically unrecognized or currently extinct hominoids leading to the latter's disappearance, or even that any of the images do so. Nevertheless, it is quite possible that some do reflect such "extinction encounters," whether involving other humans or nonhumans. Throughout history, human action has had the intentional or unintentional effect of rendering other groups of humans extinct or nearly extinct. At the level of different hominin species (or what are defined as such), many readers will be familiar with the debate concerning the Neanderthals (*Homo neanderthalensis*) and how far the actions of "anatomically modern humans" (*Homo sapiens*), extending their geographical range from Africa, may have been responsible, directly or indirectly, for Neanderthal extinction (e.g., Stringer and Gamble 1993; Tattersall 2002:135– 137).[1] More recently, the same question has been raised in regard to the little hominins called *Homo floresiensis* (Morwood and Van Oosterzee 2007; Forth 2008:282), who for some time probably shared Flores Island with *Homo sapiens*. Far earlier, Tobias suggested, the robust Australopithecines (now called *Paranthropus*) may have suffered what has been called "genocidal extinction" (Tobias 1965:32–33; see also Landau 1991:165) at the hands of members of more gracile but contemporary hominin species—ones more closely linked to the line leading to *Homo sapiens*. And, in mainland Southeast Asia, it is similarly possible that another hominin, *Homo erectus,* played a significant part in the disappearance of the huge ape *Gigantopithecus*, through either hunting or competitive exclusion (see Ciochon et al. 1990:197–214). But scenarios such as these are not only a product of paleoanthropological theorizing. For the theme of extinction brought about by invading humans occurs not only in legends told on Flores Island, but also in oral traditions from Taiwan, Sri Lanka, and Nepal which describe how, in the past, people set out to exterminate what local narratives describe as neighboring populations of wildmen.

Anthropology, Ethnology, and the Wildman

If for no other reason, such non-Western images of wildmen should be of interest to anthropologists for the ways they may illuminate the possible folk roots of scientific thought on human evolution, a matter to which I return below. But the wildman holds other sorts of interest for anthropology, including cultural anthropology or ethnology, and particularly areas like ethnozoology or folk zoology (the study of vernacular systems of knowledge concerning nonhuman animals) and cognitive anthropology.

To begin with, wildman categories raise questions for the study of what can broadly be called classification, including ethnozoological classification. A major question is what kind of entity, category, or thing—empirical or nonempirical—are (or were) the hairy hominoids that people, in several different parts of the world, speak of and often represent as extinct or becoming extinct? To the extent that anthropologists and other scholars have given them any attention at all, wildmen have implicitly or explicitly been treated as spirits, or supernatural beings. However, as I have recently shown with reference to specific ethnographic cases (Forth 2008), wildmen are in numerous respects very different from spirits. Crucial here is the demonstration by cognitive anthropologists that, wherever people speak of spirits (gods, ghosts, demons, and the like), these are subject to a counterintuitive representation which breaches basic "ontological categories" (e.g., Boyer 2001; Atran 2002). On the one hand, spiritual beings are conceived to be like humans; they are thought to possess human intelligence, feelings, and motives. On the other, spirits are "superhuman," indeed "supernatural," being capable of performing acts (such as becoming invisible or walking through solid objects) of which humans are quite incapable.[2]

Wildmen are not at all like this; indeed, they differ ontologically from "spirits," which is to say that wildmen and spirits are typically represented in ways that reveal fundamental differences of nature or perceived "essence." However empirically unlikely they may seem as representations of existing or recently existing creatures, powers attributed to wildmen are mostly natural (or physical) rather than supernatural. Writing on the yeti, the anthropologist Fürer-Haimendorf (1954) has thus remarked that Sherpas treat this creature as just another animal inhabiting the local environment. Ethnographic evidence of this sort has led some primatologists (e.g., Reynolds 1967) to conclude that the yeti represents a scientifically undiscovered and undocumented primate species. Others have offered essentially the same interpretation for the *orang pendek* (roughly, "short man") of Sumatra (Forth 2008:149–152). While a primate (ape or monkey) is less likely to be the basis of images recorded elsewhere, local representations of wildmen reveal similar distinctions between natural and supernatural beings. Thus, the Nage of Flores insist that the extinct beings they call ebu gogo were not anything like "spirits" (*nitu*). In much the same vein, Colarusso (1980:258) observes that wildmen known to rural people of the Caucasus do not appear as characters in fables or fairy tales, and that local Caucasians generally speak of them as "mundane beasts." Whatever they are, wildmen in many parts of the world thus sound, more than anything else, like zoologically plausible creatures. To be sure, some wildmen are credited with fantastic powers. For example, the *umang*, a kind of hairy hominoid recognized by the Karo people of northern Sumatra, can function as familiars of spirit mediums or shamans (Steedly 1993), and a similar significance is ascribed in Madagascar to small wildmen called *kalanoro* (Sharp 1993). Yet it is important to note that extraordinary, nonempirical powers are also attributed, in many of the world's cultures, to fully attested natural species (for example, cats, as the familiars of European witches). In fact, on the whole, known animals are far more fantastically represented—in myths and folktales and in religious or mystical practices—than are wildmen.

As images or categories, therefore, wildmen are qualitatively distinct from spiritual beings, and for this reason require a quite different kind of study (Forth 2005). As components of local systems of classification which further comprise animals recognized by modern science, the categories hold a particular interest for ethnozoologists. Like several named kinds of hairy hominoids recognized or postulated by people in other parts of Flores Island, the reputedly extinct ebu gogo are described by the Nage as "intermediate" between humans and animals. It is important to recognize that the Nage language has separate terms denoting categories that correspond to vernacular English "human" (*kita ata*) and "animal" (*ana wa;* Forth 2004, 2009). The Nage also employ distinct numeral classifiers—words used when enumerating instances of a category—when speaking about humans and animals. Animals are thus counted as *eko*, "tails," and humans as *ga'e* ("masters," but in this context to be understood as "persons"). Nage usually enumerate the ebu gogo also with "ga'e." But in accordance with the intermediate status of these creatures in relation to the animal/human contrast, people explain that this usage simply reflects their general hominoidal form and erect posture, and does not mean that the ebu gogo were fully human, like themselves. In spite of a relativist view that has gained wide acceptance in cultural anthropology (see, e.g., Viveiros de Castro 1998), ethnography lends considerable support to the cognitive universality of the distinction between human and animal.

Not only is the binary contrast widespread (and discernible even where no single local term denotes nonhuman animals), but studies of ethnobiological classification suggest that people the world over define humans and animals in much the same way. Perhaps the major criterion of zoological humanity recognized by humans generally is the possession of articulate language. Others include the production of material culture (including, of course, technology) and a practice of material exchange.

As an exclusive and definitive criterion of humanity, the ability to manufacture as well as use tools is somewhat contentious. Chimpanzees appear to produce tools and may even display regional variation in the way they do so (McGrew 2004). Nevertheless, the general point is that non-Western people typically regard human beings as something quite distinct from animals—even though in certain situations they may represent animals (or spirits manifesting in animal form) as something like human beings in disguise, or as having a hidden existence as people—and in this connection will regularly cite language and culture as exclusively human traits. Of course, partly on these grounds and usually as a result of relative ignorance, humans on first encountering previously unknown humans have sometimes mistaken them for nonhumans, or subhumans. Contrariwise, in certain historical circumstances, people have apparently mistaken animals, specifically anthropoid apes, for human beings, as for example when early European travelers mistook newly encountered gorillas and chimpanzees for humans.[3] It is similarly undeniable that humans sometimes speak of particular groups or categories of other humans as being more like animals, or less human, than themselves. Far from providing evidence for culturally various conceptions of the human/animal boundary, or radically different ontologies, however, this kind of representation is better understood as manifesting a special ideological or "symbolic" cognition, which one confuses with empirical, or "encyclopedic," knowledge (Sperber 1975) at one's analytical and philosophical peril.[4] Ultimately, and in the absence of zoological specimens, one cannot be absolutely sure that the ebu gogo wildmen of Flores, for example, do not exemplify a symbolic representation, or did not do so at one time. In other words, this Florenese image may have originated as a sort of ethnic slander perpetrated against some other human group. On the other hand, what the referent of this derogatory image may have been is anything but clear (see Forth 2008:275–280). And casting further doubt on the hypothesis are the very different ways in which Florenese people nowadays represent disliked human others, whom they do not describe as physically distinct but only as behaviorally and morally different from themselves.

The Florenese conception of wildmen as creatures intermediate between humans and animals does not of course prove that the images lack any basis in real living beings, including ones that may be either fully human or non-human. Indeed, with reasonable accuracy one could describe the great apes, as well as fossil hominins like *Australopithecus africanus,* as intermediate between *Homo sapiens* and lesser nonhuman primates. In one respect, however, "intermediate" is not the best term, and it would be better to say that the ebu gogo wildmen, for example, are spoken of as locally extinct beings whose ontological status, as human or animal, is uncertain or ambiguous. Florenese people, it should be emphasized, do not regard the ebu gogo as intermediate (between humans and nonhuman animals) in any evolutionary sense. Their ideas of connections among different zoological species do not approach anything like a Darwinian view, and by the same token they do not conceive of locally recognized hominoids as ancestors or forerunners of human beings, local or otherwise (Forth 2006). If the ebu gogo were a thoroughly imaginary representation, symbolically constructed for some ideological end, the mixture of human and animal traits could have been taken much further—for example, they could have been given tails, horns, or fangs. In fact, they lack all of these attributes, and besides their reputedly hairy and shorter bodies and somewhat simian facial features, the main traits contradicting their full humanity are negative: the reputed lack of a material culture and a fully articulate language. On the other hand, if the ebu gogo image were taken as reflecting empirical beings that lacked (or appeared to lack) proper language and culture, it would have to refer either to human beings who were technologically extremely primitive (indeed, more primitive than anything known to modern ethnography) or a non-*sapiens* species of the genus *Homo.* If we allow for exaggeration in the other direction, which is to say an anthropomorphization of an animal species, another possibility might be some sort of ape. But while orangutans (*Pongo pygmaeus*) are found on Sumatra and Borneo, Flores and other eastern Indonesian islands lack known apes. Nor is there evidence for any in recent geological times. The only nonhuman primate on Flores, introduced by human immigrants a few thousand years ago, is a monkey, the long-tailed macaque (*Macaca fascicularis*).

Stories of Extermination and the Moral Dimension of the Human/Animal Opposition

The classificatory contrast of human and nonhuman has an equally universal moral dimension, for everywhere it relates to how humans should treat creatures considered nonhuman. Recalling earlier European understandings of great apes as types of "men," indeed "wild men," some geneticists have recently argued that chimpanzees and *Homo sapiens* should be placed in one and the same genus (Watson et al. 2001). Regardless of how well supported this reclassification may be on scientific grounds, especially interesting in the present context is the way this taxonomic argument has been linked with a moral argument that something like human rights should be extended to the great apes (see Cavalieri and Singer 1993).

To say that humans and nonhumans are everywhere valued differently is certainly not to suggest that humans do not kill or harm fellow humans—that is, beings they identify, ontologically if not ideologically, as human beings—or that they do not often do so with impunity, most notably in the context of warfare. Nevertheless, especially where they concern members of the own group—which category, I would argue, people can always extend to include humans everywhere (see Forth 2009)—the moral implications of killing humans and animals are quite different. In this respect too, wildmen can be called "intermediate"; for just as they present a classificatory ambiguity, so they present a moral one. Noteworthy in this respect are reports by Europeans who claim to have encountered specimens of locally recognized hairy hominoids (for example, in Sumatra and Central Africa) but who, although armed, found themselves unable to fire on the creatures because of their too human appearance.[5] Especially in contexts of warfare and territorial competition, on the other hand, participants in other cultural traditions have evidently had no compunction in killing wildmen, as narrative traditions concerning putative encounters with such creatures would especially suggest. Not only do such stories provide a major source of information on human/wildman relations and on local knowledge regarding how the hairy hominoids became extinct or their numbers significantly reduced, they also bear on what may be called their ecological plausibility. The Nage legend of the ebu gogo serves as a useful illustration.

The Nage people occupy an upland region in the central interior of Flores Island. Their wildman legend more particularly concerns a local group named 'Ua. According to Nage oral history and clan genealogies, the 'Ua people moved into what is their present territory between two and three centuries ago. Previous to this the territory, a densely forested area high on the slopes of a volcano, was uninhabited and unclaimed by any current Nage group. A decade or two after settling in the region, the 'Ua folk began to encounter a physically distinct population of hirsute, dull-witted, and apparently cultureless hominoids living in a cave higher up the mountain. These were the ebu gogo. The relationship between the 'Ua people and the ebu gogo was at first neutral; it was not beneficial, at least not to the 'Ua folk, since it involved no positive reciprocity (as one should expect if both groups were *Homo sapiens*). However, with the expansion of human agricultural settlements in this previously uninhabited region, the hominoids became more and more a threat to 'Ua subsistence, as they began stealing produce from cultivated fields. Finally, the 'Ua people decided to exterminate the ebu gogo. This they did by confining them inside their cave and igniting a great fire inside the entrance. The creatures were all killed in the resulting conflagration, a reputed event that can accurately be called a holocaust.

This is necessarily a condensed summary of the tale. (Further details can be found in Forth 2008, chapter 2.) By the same token, it may err somewhat on the side of naturalism. There are to be sure evidently fantastic touches— like a description of how the gluttonous hominoids first attended an 'Ua feast, where they drank to excess and swallowed eating and drinking vessels as well as food and palm wine. The creatures are also described as speaking, albeit in a muddled and imperfect way. Yet another evidently imaginative component is the escape, mentioned in some versions of the tradition, of two opposite-sex specimens that were absent during the conflagration and managed to flee to another part of Flores Island—thus, as it were, facilitating a Hollywood-style sequel. The sequel, however, does not exist, and all Nage people agree that the ebu gogo are now extinct, at least in the Nage region.

Apparent narrative devices aside, the story of the ebu gogo appears basically to concern two populations, one human and the other subhuman (or ambiguously human), which come to compete over subsistence resources in a common territory. In consequence, the humans conspire to destroy the subhuman ebu gogo, and this they accomplish through the use of fire. Nage say there were no ebu gogo living elsewhere when they were encountered by the 'Ua people. This might suggest a remnant population, surviving in just one or a few isolated locations long after Florenese humans had come to occupy less elevated parts of Flores Island, including areas closer to the coast. There are differently named and somewhat differently described categories of similarly small hairy hominoids reported by other ethnolinguistic groups from other parts of Flores. In addition, stories of such creatures being killed by fire—although not totally exterminated by this means—are found in other parts of central Flores. Apart from Flores, the holocaust theme occurs only in Sri Lanka, among traditional hunter-gatherers collectively known as Veddas. Like the Nage story of ebu gogo, this Sri Lankan legend—concerning a kind of hairy hominoidal creature named *nittaewo*—moreover refers to a single specific location in eastern Sri Lanka (Forth 2008:182–187). Somewhat similar traditions recounting the disappearance of small humans or hominoids also occur on the large Indonesian island of Sulawesi and on Taiwan, but these do not include the theme of destruction by fire. Nor do Sherpa folktales, which relate the destruction of groups of yeti, but evidently not the elimination of the entire "species" so named (192–193).

The occurrence of similar tales in other parts of Flores and in at least one part of South Asia may cast doubt on the historicity of the ebu gogo legend. It could, for example, indicate that the stories merely represent instances of a common mythical tradition. On the other hand, it should be remarked that the destruction of wildmen by fire is not a widespread motif in world folklore, although stories involving the killing of other creatures by this means are more common (see Arne 1973; Thompson 1975). Also, since wildmen are generally described as much stronger and tougher than humans, the use of fire—a well-developed hunting technology in Flores and elsewhere—could suggest a practical means of combating a whole group of troublesome creatures employed in similar ecological circumstances. For example, in the Rawe district of central Flores, populations of chicken-stealing wild cats have been reduced in recent decades by firing rock cavities in which the animals typically take refuge (Forth fieldnotes 2010). Within Indonesia, the theme of exterminating wildmen by burning appears confined to Flores, so variants from several parts of that island could in fact reflect a story, possibly grounded in empirical events or a series of events, that has spread from one part of the island to others. Rather than deliberate attempts at wholesale extermination, these events could have formed a pattern of interaction between humans and human or nonhuman others which, in the relatively distant past and unfolding over a long period, and as part of an ongoing competition over territory and resources, eventually resulted in the wildmen's extinction.

A recent review of evidence from various parts of the world has shown that stories recounting the capture or confinement of wildmen are far more commonly encountered than are stories of burning or otherwise exterminating them (Forth 2008). Confinement, if not capture, is of course an element of the Nage story of the ebu gogo's extermination, as the creatures were initially trapped inside a cave. Elsewhere, the theme of wildman capture occurs in the European mediaeval motif of the wildman hunt (Bernheimer 1952:25), a motif that recalls eastern Indonesian stories about wildmen recently encountered by hunters pursuing other creatures (e.g., Forth 2008:73–75, 92). Other stories of captured specimens—or what can sometimes more accurately be called rumors—are found in reports from Sumatra and the eastern Indonesian islands of Sumba and Sumbawa; in my own ethnography of various regions of Flores (Forth 2006, 2008:110–113); and in accounts, including some famous in cryptozoological circles, of hairy hominoids allegedly captured and confined in Vietnam, the Caucasus, western Canada, Polynesia, and Madagascar.

One example of such tales is a story concerning the mid-twentieth century capture of a hairy wildman inside a cave in the mountains of Vietnam. This is recorded in one of several books on the Vietnam War by the Australian journalist and war critic Wilfred Burchett (1965; see also Forth 2008:169–170), who accepted the story as genuine. The creature was apprehended, bound, and taken to a guerilla camp, but due to its rapid decline, its captors later decided to return it to its cave. On the way to the cave, the creature died and was buried where it expired. Now thoroughly assimilated into sasquatch or bigfoot lore, a news report from 1884 concerns a small hairy hominoid captured in that year near the frontier town of Yale in British Columbia (Forth 2007, 2008:211–215). Interestingly, the Canadian news story is in several respects reminiscent of a report of an 1879 capture of a similar creature in northern Madagascar which found its way into the *Proceedings of the Royal Geographical Society* (Ransome 1889; see also Sibree 1896). The Madagascar story appears connected with Malagasy representations of small hominoids named kalanoro, which local people say used to be found on the island but are now extinct (G. Sodikoff, personal communication). Also deriving from the mid- to late nineteenth century is the legend of a large "wild woman" named Zana, an individual evidently much larger than the Vietnamese, Canadian, and Malagasy captives, all of whom were male, who was captured in the Caucasus. As in several stories from Southeast Asia, Zana is supposed to have produced offspring by a human male. Curiously, although it relates to a much smaller wild woman, a legend I recorded in 2005 on the eastern Indonesian island of Sumba (Forth 2008:100–101) is remarkably similar to the story of Zana and is also set in the nineteenth century.

Admittedly, none of these reports has ever been substantiated empirically. It is also noteworthy that almost as common a feature of wildman images as victims of capture is their representation as captors of humans, and especially as abductors of children. Even so, it is quite clear that humans in several culturally and historically very diverse places have, on reputedly encountering wildmen, attempted to pursue and capture them, sometimes with the aim of killing them. This is perhaps not so extraordinary when one considers the lengths to which Westerners have gone to acquire specimens of every zoological kind. One thinks, for example, of Alfred Wallace's "relentless pursuit" of Bornean orangutans, resulting in the killing of 15 specimens in six months (Raby 2001:107). The example is perhaps especially apt, as the ultimately Malay name of these creatures, *orang utan,* translates as "man (person) of the woods," or even "wild man." As is well known, orangutans will almost certainly become extinct in the next several decades, unless, ironically, they are able to survive in zoological collections. In the not-too-distant future, therefore, humans will be able to relate to these red-haired "wild men" only as empirically fully attested captives, and perhaps with some good scientific reason. Yet the more general human propensity for pursuit and capture acquires an additional significance in the light of non-Western representations of human/wildman relations that are not so obviously linked with the objectives of zoological investigation.

The Wildman and the Evolution of Humans and Human Culture

Capture and confinement of animals, leading to domestication, is of course a long-standing process of human cultural evolution. According to a view commonly rehearsed in textbooks of anthropology, it is by means of culture that humans have domesticated themselves, thus actively participating in their own physical evolution. The study of human physical evolution bears in another way on the wildman. Without too much stretching of definitions, paleoanthropology and archeology can be said to provide evidence for the actual existence of wildmen, or what—if they could be encountered in the flesh—would surely pass as such. I refer to the extinct pre-*sapiens* hominins that were ancestral or related collaterally to anatomically modern humans. Yet another kind of anthropological study, one focusing on Western cultural images, raises the question of how far another sort of wildman representation, one evidently derived from prescientific sources, may have contributed to paleoanthropological reconstruction and theories of human evolution. For example, archeologist-philosopher Wiktor Stoczkowski (2002) has identified the European wildman, particularly the figure of late mediaeval art and literature, as an important (albeit implicit and unrecognized) source of many Western scientific ideas concerning pre-*sapiens* members of the genus *Homo*.

Stoczkowski's interpretation appears thoroughly reasonable. Yet it raises several questions, including how precisely the mediaeval figure, now mostly known only to historians and literary specialists, has remained imprinted in modern thought and has survived in the models and reconstructions of archeologists and biological anthropologists. In linking the European wildman with archaic hominins, Stoczkowski moreover refers mostly to hypothetical behavioral and psychological aspects of ancient hominin existence, and not to their physical form, which of course has a more solid empirical grounding in fossil evidence. Similarly, Misia Landau asks how far "preexisting narrative structures" may have contributed to an interpretation of the fossil record, a task which, she notes, would be much more burdensome without recourse to these structures (1991:149). Among the latter, one finds themes of heroic struggle against lesser beings that are consequently rendered extinct. However, the traditions to which Landau refers all derive from Western culture. Like Stoczkowski and others, Landau does not take into account the very comparable narratives concerning wildmen found in non-Western cultures, such as those told by the Nage of Flores and other rural Southeast Asians.

If cultural representations of wildmen, their relations with humans, and ideas concerning the human hunting, capture, and killing of such creatures are not exclusively Western—as clearly they are not—then we are left with an intriguing problem, namely, what could account for the widespread occurrence of such representations? Since explaining wildmen as the purely imaginary product of one cultural tradition (e.g., the Western) is an obvious nonstarter, we might interpret the images as equally fantastic products of a number of separate and quite various cultures. Yet this approach only raises the question of why the often striking similarities appear. One possibility is that the images and traditions reflect an equally widespread experience of something empirical. This "something," however, need not be the same thing in every case, nor an entity which in every case is reflected with equal accuracy in the images. Thus, while the orang pendek (or "short man") of southern Sumatra could very well reflect experiences of a locally rare and unfamiliar ape, the western North American sasquatch is hardly likely to be an undiscovered primate. In fact, if the sasquatch is not largely or entirely imaginary, it would have to be what "believers" say it is, namely, a giant hairy hominoid. The idea that all wildman images—or indeed any of them—reflect relatively recent experiences or cultural memories of once contemporary non-*sapiens* hominins may seriously strain credulity. Yet, as has been demonstrated elsewhere (Forth 2008, chapter 10), such an interpretation is significantly more plausible for certain parts of the world, including Flores Island, than for others. Representations, mostly distant in space or time from actual zoological referents, of such "near humans" as anthropoid apes might therefore seem a more likely source of wildman images. Yet this explanation, too, fits some instances—for example, wildman images from Africa and parts of western Southeast Asia—far better than it does others. Wildman images from around the world, therefore, are unlikely to be of a single ontological or epistemological piece.

Not inconsistent with the foregoing is another possibility: representations of wildmen in different cultures may be understood as similar (though not of course identical) responses to an archetype, a pan-human proclivity to imagine humanlike beings distinguished from humans by the possession of a minimum of nonhuman or animal traits (see Heuvelmans 1990). This proclivity may explain why, in an earlier day, Europeans considered the great apes as "wild men" (that is, as more human than most primatologists would currently conceive them to be)—and, indeed, like "wildmen" as imagined in a variety of the world's cultural traditions. Particularly interesting in this connection is a series of quite specific similarities between wildman figures, demonstrably traceable to experience of real apes, which are represented by Africans and Southeast Asians occupying territories adjacent to or overlapping with regions occupied, or recently occupied, by orangutans, chimpanzees, and gorillas (Forth 2008).[6] What this could more specifically suggest is that the images are partly shaped by a sort of "cognitive distance" between local people and unfamiliar and rarely encountered objects of their observations. Alternatively, it could involve empirical beings known only secondhand, through stories passed over several generations. In either case, what could account for this distance is, if not the complete extinction of a species or population, then their local extirpation, resulting either from killing (hunting or the deliberate destruction of bothersome or notionally dangerous creatures) or competitive exclusion from shared environments.

Summary and Conclusions

Whatever the ontological status of particular wildmen—that is, whatever basis the hairy hominoids may or may not have in zoological reality—the figure of the wildman provides a useful way of thinking about things; it is, in Lévi-Strauss's famous phrase, "good to think." Among these are our understanding of human evolution and its further possible course, and human tendencies toward the extermination of other species as well as ethnocide or genocide. As a commonly encountered image seemingly comparable to representations of zoologically attested natural beings, the wildman is quite distinct from spiritual beings. Although not clearly universal in the same way as spirits, the wildman is nonetheless a widespread representation that is not specific to a single cultural tradition or kind of social system, and so cannot readily be explained as an imaginary social construct or a special cultural symbolism. Stories of wildmen, including legends of their destruction and extinction, reflect an equally widespread human attitude toward other zoological kinds, and perhaps especially to creatures that appear hominoid but not quite human and which may therefore challenge the conceptual boundary of humanity and animality. Regularly revealing themes of capture, confinement, and extermination, scenarios such as the one depicted in the Florenese story of the ebu gogo, like comparable traditions from other parts of Southeast Asia and elsewhere, reveal that the idea, if not always the reality, of a deliberate extermination of entire groups is not a recent invention of modern (or modernizing) societies. It might even be a significant factor of human evolution. If there is an archetype to which various wildman images respond, one of its main components appears to be exclusion, persecution, and extirpation leading to extinction, either as the fate of the wildman or as an aspect of relations with humans.

NOTES

1. Some paleontological evidence has been interpreted as indicating not only killing of Neanderthals by modern humans but possible consumption, or "cannibalism," of them as well (Rozzi et al. 2009).

2. Normal humans, at any rate, are incapable of such acts. In some societies, shamans, for example, are credited with similar superhuman powers and are therefore subject to the same counterintuitive representation as spirits.

3. The Englishman Andrew Battell, who found himself in Central Africa around the beginning of the seventeenth century, is one example. Battell described what were very probably gorillas as differing in their physical form hardly at all "from a man" (Ravenstein 1901:54). Even though he noted that the creatures could not speak, his corollary statement that they "have no more understanding than a beast" implies that he did not fully regard them as "animals" either.

4. An example of the distinction concerns the Nage representation of malevolent spirits as killing human beings in the form of sacrificial water buffalo. Conversely, in certain contexts Nage speak of these spirits, who in their own domain live exactly like humans, as being susceptible to slaughter when they, the Nage, sacrifice buffalo (Forth 1998). This representation, however, is relevant only to religious ritual and myth, and one would be mistaken to infer from it that Nage do not ordinarily distinguish humans, animals, and spirits, or that they always identify water buffalo with anthropomorphous spirits. Outside of sacrificial ritual, they treat water buffalo like any other domestic animal.

5. Probably the best-known instance is the 1924 account by Van Herwaarden, a Dutch timber prospector, of his reputed encounter with a female specimen of the Sumatran hairy hominoid usually known as "orang pendek." Similarly, a European named Pessina, hunting in the Masisi forest near Lake Kivu in the Congo (DRC) in 1963, claimed to have seen coming toward him what he first thought was a chimpanzee. Much like Van Herwaarden, he noticed something human in its form and behavior and therefore hesitated before taking a shot, thus giving the creature sufficient time to escape (Cordier 1973:188). A non-European instance of the same ambivalence is provided by David Labang, a Malaysian park ranger who reported observing what he described as a small, hairy, naked man in Pahang, in Peninsular Malaysia. Labang too "wavered between shooting and capturing the creature, but decided on the latter." He chased after the hominoid, but it proved too agile and escaped (McNeely and Wachtel 1988:261).

6. Particular instances concern the Sumatran hominoid called orang pendek, reported from a region where a remnant population of orangutans may survive or

survived until a few hundred years ago, and mostly similar but differently sized hominoids reported from Central Africa, in or near regions occupied by chimpanzees, bonobos, and gorillas. Features shared by the African and Sumatran images include hair falling over the forehead or the eyes (a feature also mentioned for the Himalayan yeti); an association with herds of wild pigs; feeding on freshwater crustaceans, larvae from rotten trunks, and ginger plants; stealing fish; and disturbing people spending the night in temporary forest shelters.

REFERENCES

Arne, Antti
  1973 The Types of the Folktale: A Classification and Bibliography. Translated and enlarged by Stith Thompson. 2nd rev. Helsinki: Academia Scientiarum Fennica.

Atran, Scott
  2002 In Gods We Trust: The Evolutionary Landscape of Religion. New York: Oxford University Press.

Bernheimer, Richard
  1952 Wild Men in the Middle Ages: A Study in Art, Sentiment, and Demonology. Cambridge, MA: Harvard University Press.

Boyer, Pascal
  2001 Religion Explained: The Evolutionary Origins of Religious Thought. New York: Basic Books.

Burchett, Wilfred G.
  1965 La Seconde Résistance: Vietnam 1965. Michel Deutsch, trans. Paris: Gallimard.

Cavalieri, Paola, and Peter Singer, eds.
  1993 The Great Ape Project: Equality beyond Humanity. London: Fourth Estate.

Ciochon, Russell L., John W. Olsen, and Jamie James
  1990 Other Origins: The Search for the Giant Ape in Human Prehistory. New York: Bantam Books.

Colarusso, John
  1980 Ethnographic Information on a Wild Man of the Caucasus. In Manlike Monsters on Trial: Early Records and Modern Evidence. Marjorie Halpin and Michael Ames, eds. Pp. 255–264. Vancouver: University of British Columbia Press.

Cordier, Charles
  1973 Animaux Inconnus au Congo. Zoo 38(4):185–191.

Forth, Gregory
  1998 Beneath the Volcano: Religion, Cosmology and Spirit Classification
  among the Nage of Eastern Indonesia. Leiden: KITLV Press.
  2004 The Category of "Animal" in Eastern Indonesia. Journal of Ethnobiol-
  ogy 24(1):51–73.
  2005 Hominids, Hairy Hominoids and the Science of Humanity. Anthropology Today 21(3):13–17.
  2006 Flores after Floresiensis: Implications of Local Reaction to Recent Pal-
  aeoanthropological Discoveries on an Eastern Indonesian Island. Bijdragen
  tot de Taal-, Land- en Volkenkunde 162(2–3):332–345.
  2007 Images of the Wildman inside and outside Europe. Folklore 118:261–
  282.
  2008 Images of the Wildman in Southeast Asia: An Anthropological Perspective. London: Routledge.
  2009 Human Beings and Other People: Classification of Human Groups and
  Categories among the Nage of Flores (Eastern Indonesia). Bijdragen tot de
  Taal-, Land- en Volkenkunde 165(4):493–514.

Fürer-Haimendorf, C. Von
  1954 The "Scalp" of an "Abominable Snowman"? A Yeti-Hide Head-Dress.
  The Illustrated London News 224:477 (March 27).

Heuvelmans, Bernard
  1990 The Metamorphosis of Unknown Animals into Fabulous Beasts and of
  Fabulous Beasts into Known Animals. Cryptozoology 9:1–12.

Landau, Misia
  1991 Narratives of Human Evolution. New Haven: Yale University Press.

McGrew, William
  2004 The Cultured Chimpanzee: Reflections on Cultural Primatology. Cambridge: Cambridge University Press.

McNeely, Jeffrey A., and Paul Spencer Wachtel
  1988 Soul of the Tiger: Searching for Nature's Answers in Exotic Southeast
  Asia. New York: Doubleday.

Morwood, Mike, and Penny Van Oosterzee
  2007 A New Human: The Startling Discovery and Strange Story of the "Hob-
  bits" of Flores, Indonesia. New York: HarperCollins.

Raby, Peter
  2001 Alfred Russel Wallace: A Life. Princeton: Princeton University Press.

Ransome, L. H.
1889 The River Antanambalana, Madagascar. Proceedings of the Royal Geographical Society and Monthly Record of Geography 11(5):295–305.

Ravenstein, Ernest G., ed.
1901 The Strange Adventures of Andrew Battell of Leigh, in Angola and the Adjoining Regions. London: Hakluyt Society.

Reynolds, Vernon
1967 The Apes: The Gorilla, Chimpanzee, Orangutan, and Gibbon; Their History and Their World. New York: E. P. Dutton.

Rozzi, Fernando V. Ramirez, Francesco d'Errico, Marian Vanhaeren,
Pieter M. Grootes, Bertrand Kerautret, and Véronique Dujardin
2009 Cutmarked Human Remains Bearing Neanderthal Features and Modern Human Remains Associated with the Aurignacian at Les Rois. Journal of Anthropological Sciences 87:153–185.

Sharp, Lesley
1993 The Possessed and the Dispossessed: Spirits, Identity, and Power in a Madagascar Migrant Town. Berkeley: University of California Press.

Sibree, James
1896 Madagascar before the Conquest: The Island, the Country, and the People. London: T. Fisher Unwin.

Sperber, Dan
1975 Rethinking Symbolism. Alice L. Morton, trans. Cambridge: Cambridge University Press.

Steedly, Mary M.
1993 Hanging without a Rope: Narrative Experience in Colonial and Postcolonial Karoland. Princeton: Princeton University Press.

Stoczkowski, Wiktor
2002 Explaining Human Origins: Myth, Imagination and Conjecture. Cambridge: Cambridge University Press.

Stringer, Christopher, and Clive Gamble
1993 In Search of the Neanderthals: Solving the Puzzle of Human Origins. New York: Thames and Hudson.

Tattersall, Ian
2002 The Monkey in the Mirror: Essays on the Science of What Makes Us Human. New York: Harcourt.

Thompson, Stith
1975 Motif-Index of Folk-Literature: A Classification of Narrative Elements in Folktales, Ballads, Myths, Fables, Mediaeval Romances, Exempla, Fabliaux, Jestbooks, and Local Legends. Revised and enlarged edition. 5 vols. Bloomington: Indiana University Press.

Tobias, Phillip V.
1965 Early Man in East Africa. Science 149:22–33.

Van Herwaarden, J.
1924 Een Ontmoeting met een Aapmensch. De Tropische Natuur 13:103–106.

Viveiros de Castro, Eduardo
1998 Cosmological Deixis and Amerindian Perspectivism. Journal of the Royal Anthropological Institute 4(3):469–488.

Watson, Elizabeth E., Simon Easteal, and David Penny
2001 Homo Genus: A Review of the Classification of Humans and the Great Apes. In Humanity from African Naissance to Coming Millennia: Colloquia in Human Biology and Palaeoanthropology. Phillip V. Tobias et al., eds. Pp. 307–318. Johannesburg: Witwatersrand University Press.

EPILOGUE

PROLEGOMENON FOR A NEW TOTEMISM

Peter M. Whiteley

Totemism in the Imagination of Nature

> By endowing nature with social properties, humans are
> doing more than granting her anthropomorphic attributes,
> they are mentally socializing the relationship they establish
> with her . . . In order to exploit nature, humans weave a
> network of social relationships between themselves, and it is
> most often the form of these relationships that provides the
> conceptual model for their relationship with nature.
>
> —PHILIPPE DESCOLA, *IN THE SOCIETY OF NATURE*

Nature, natural species, as Lévi-Strauss taught, are *bonnes à penser,* good to think.
It is through conceptualizations of nature and their usage as metaphors that we
have imagined ourselves as humans, in our groupings, in our actions, and in terms
of our social reproduction. Among the Hopi, a Native people of northern Arizona,
for example, the imagination of social difference is predicated upon observation
of and participation in the differentiated and classified natural environment. Like
many small-scale, non-Western societies, Hopis divide their social order into to-
temically named entities—matrilineal clans—like Bear, Sun, Parrot, Spider, Rattle-
snake, and Sparrowhawk (e.g., Bradfield 1973). Unlike some other Amerindian
societies (as on the Northwest Coast), Hopi clans do not consider themselves to
be descended from natural species and entities, but there is a sense of mystical coparticipation: the clan's *naatoyla,* emblem, is also its *wu'ya,* "ancient." Moreover, in
ceremonies like the Snake Dance, in which ritual initiates dance with live snakes,
they enact that coparticipation not for symbolic purposes only, but because they
actively believe this will summon the clouds and their rainfall.

While such totemic relationships (to use this term broadly) between culture and nature have their locus classicus in small-scale indigenous societies (e.g., Willis 1974), totemist perspectives are present in all forms of human society. Edmund Leach (e.g., 2001) reminded us that implications of totemic thought in British culture occur in "animal categories and verbal abuse," and in pervasive discursive correlations between thoroughbred horse breeding and aristocratic distinctions. Or, to take another domain, consider "trees," "roots," and "forests" in contemporary cladistics: even the most sophisticated scientific explanations are metaphorically framed within vernacular metaphors of natural processes. It seems plausible that these types of conceptual vehicles of association and explanation are virtually hard-wired into human cognition.

The current conceit that seeking to preserve natural forms is both a new idea and the contribution only of enlightened Western liberal democracies is palpably untrue, as Sodikoff's discussion of Malagasy *fady* ("taboos") clearly shows. And the Hopi case may be helpful here also: Hopi hunting taboos, as Beaglehole described seven decades ago, are designed to exert a conservationist effect:

> To understand the use of ritual as an aid towards conservation, it may be recalled that the Hopi attitude towards animals, like that of all the Pueblo peoples, is one of respect and esteem. Animals may not be ruthlessly destroyed or wantonly exploited just for the love or excitement of the chase. They must be protected, entreated humbly not to become angry if killed, and urged to give themselves or their young for the use of their human kinsmen. Taken in conjunction with the fact that prayer sticks are placed on shrines or buried in fields during the winter solstice to ensure fertility of all animals, whether wild or domesticated, it is evident that [the] propitiatory aspect of ritual serves to preserve animal life for continued use by checking evil results that would inevitably follow from uncontrolled carelessness, neglect, ill-treatment, or the operation of obscure other-worldly forces ... [Hopi hunting ritual is used] to help secure success and to preserve the fauna of the environment from thoughtless exploitation. (Beaglehole 1936:22–24)

So the claim that we moderns have developed something distinctive and superior to older views of environmental practice by indigenous communities is specious and neglects that such communities, as a general rule, have had virtually zero impact on the present species extinction threat. And in seeking to adopt favored contemporary indigenous communities as mascots for sustainable development—as by the recent United Nations Millennium Development Goals Equator Awards presented in September 2010 (e.g., United Nations Development Programme 2010)—our intention seems less to learn from them, and certainly not to treat them as autonomous agents or sovereign sociopolitical entities, than to project them yet again as archaic survivals, or neo–noble savages, as Gerald Vizenor (e.g., 1999) has it.

Likening the disappearance of species to the disappearance of the cultures and languages of comparatively small indigenous societies (as Perley, chapter 6, argues), or even of creoles and pidgins (as Garrett, chapter 7, argues), is thus an extension of preexisting human cognitive propensities in natural metaphor. Attempting to engage a moral force in the problem of extinctions is clearly also a projection of our own individual sense of mortality: indeed, the consciousness of death serves as a key, as Freud among others emphasized, to what distinguishes us from other earthly species. The foundation of concerns with all forms of extinction is Thanatos, ever-present to our consciousness, particularly as we grow older.

To suggest that this projection is a natural part of the human condition is not to deny its timeliness and value in the present iteration. But there are dangers, especially those that continue the old equation of Natives with Nature which has been with us at least since 1580, with Montaigne's essay "Of Cannibals" (1993). The current transformation may offer no more sophisticated solution to problems either natural or cultural. Extinction of indigeneity is also not new: the idea of the "Vanishing Indian" has long served the West as an elegiac metaphor of Arcadian extinction—the vanishing somehow ideologically removable from colonial displacement, introduced disease, war, and nation-state hegemony. And those hegemonic forces strongly persist in Native American life, as elsewhere. At the American Anthropological Association meetings in 1986, Native linguist Ofelia Zepeda (Tohono-O'odham) proposed a resolution to get endangered indigenous cultures and languages included with endangered species—since this seemed to be the only way to attract attention to the problems facing indigenous languages. This equation evoked some moral horror among the participants who voted the resolution down, and it should continue to do so today. But to question its aptness or force in effecting change requires some examination of its conceptual underpinnings.

To put this another way, if morality is projected as the bridge across the culture/nature divide that may save species, languages, and traditional social systems and cultures, we may want to inspect the columns that this new totemist bridge rests upon. If we accept Lévi-Strauss's (e.g., 1969) basic arguments, the realization of culture out of nature is a universal process of human social cognition and imagination. Identification of social differences through the metaphor of natural species differentiations may be the oldest game of culture—of the classification and ordering of experience into grids and networks of signification—in the world. And, culturally speaking, humans know that, which is why so many mythological systems show an originary wresting of cultural powers from natural forces: culture is born in agentive conflict with or distinction from nature. For the Hupa of northern California,[1] for example, the birth of culture lay in the appropriation by trickery of sacred natural powers (a special blanket decorated with pileated woodpecker scalps) by a *kixunai,* an ancestral river spirit transformed into a culture-hero, from his uncles, the Sun and Moon (Whiteley 2004:491).

In this light, contemporary concerns about competing views of nature, seen through the prism of extinctions discourse, represent a clash between old and new totemisms, thought-systems with which humans identify themselves and their values through natural metaphors. What is the difference, though, between the new totemism—which says we are lemurs or mouflons ("lemurs/mouflons are us" is perhaps more culturally fitting; see chapters 2, 3, and 8)—and the old one, in which we (Bororo) are scarlet macaws (e.g., Smith 1972), or we (Hopi clan-members) are Badgers, Spiders, and Eagles? That difference may be best captured in the opposition between so-called traditional ecological knowledge (TEK) and the increasingly hegemonic discourse of the biodiversity conservation movement, based on Western neoliberalism and technoscientific understandings of the natural world.

The difficulties of retotemizing scientific conceptions and reconstructions of species in processes of "citizenship," "repatriation," and "renaturalization" are profound. Whereas traditional social ecologies are holistic, technoscience "isolate[s] static and highly valued objects of conservation" (Heatherington, this volume). By extension, the imbrications of language use with structures of social relations and economic production (Perley and Garrett, this volume)—land, knowledge, resources, statuses, occupations—echo the holistic form for culture in general of the old totemism (as Descola [above] suggests). But in this connection, the question arises, "Language preservation or restoration for what purpose?" (see Whiteley 2003). Language restoration will not or cannot regain much for a community unless it is integrated with structures that are simultaneously restorative of social relations. Typically, what gets imagined is language-as-identity, rather than as natural means of communication: a language, like contemporary indigenist discourses that privilege "culture," as a site to justify claims to sovereign rights, becomes a symbolic projection in itself, an ostensive end rather than a directly productive social means. The conflict between linguistic approaches and language practices figures this new-old, modernist-autochthonous dichotomy of conservation interests in another dimension.

If we are to produce an efficacious critique of isolationism in this regard, we need to address its active differences from holism. We cannot effectively identify species as our social totems, in technoscientific totemism, unless we are prepared to acknowledge the interrelationality that characterizes the old totemism. As Hopi, we are only Bears *because* you are Badgers, they are Eagles, and those others are Rattlesnakes: we reproduce our society and our cosmos by *exchange* with each other—in marriage transactions, ritual prestations, and other forms of reciprocity. The natural metaphor is only meaningful because it produces or expresses a whole scheme of operational social differences: the contributions that each species, each social component, brings to the world as a whole are relative, not absolute. Charismatic species (particularly megafauna) in Western scientific thought, by contrast, are given singular, self-contained value. Perhaps that is a product and reflection of bourgeois individualism, or of the atomism—versus what we might call familism—of social relations in general that characterizes life in Western democracies.

Time and Narrative

Another critical dimension here concerns history and its absence. Both culture and nature are *agentive*, not merely passive assemblages. In this sense they are the products of history—human history for the former, natural history (though I am extending the meaning of this term) for the latter. By "history" I mean a marked and remembered cumulative series of transformative, agentive interventions in everyday life, on the one hand, and, for natural science, on the other, evolution (whether gradualist or by punctuated equilibrium). (Cultural) history is anchored in difference, even conflict, in differentiation of events from routine processes, in transformations of cultural structures that occur in the conjuncture with events. As with culture, so, after Darwin, with nature, which is seen, in the long term, as in constant flux in the creation and diachronic transformation of species and their environments.

Against these historicities, both anthropology's notion of culture and language, on the one hand, and conservationist thought's sense of natural species, on the other, tend to freeze time. Anthropology often (and notoriously) conceives of culture and language as atemporal (see Whiteley 2004), projecting an essentialized ethnographic (or linguistic) present that impedes recognizing either as constantly in process. Conservationist discourse about species appears equally atemporal. Both anthropology and conservation often effectually produce "frozen arks" as their imagined objects. The sometimes Rousseauian tendencies of activist organizations like Cultural Survival and Survival International—wanting to maintain societies in the "state of nature," if you will—mirror an equivalent stasis in the imagery of species attending much biodiversity conservation activism. And attempts at preservation and restoration that flow from these propensities thus entail a double dose of reification.

Futures?

So what hope for the future: are we compelled to imagine conservation as aiming to preserve "nature" and "culture" in a bottle? Must we now envision nature and culture (including language) only ironically, detached from unreflexive experience? Have they become for us merely symbolic capital, or social products of a postcultural and postnatural imaginary—*as if* culture and nature, so to speak? And if so, are such projections the proper object of collective human interest in conservation? In any event, do we even have a choice as to how we might imagine these interests? Culture and nature in a bottle may be our only hope, given the present and forecast rates of extinctions. We should not be too sanguine that biodiversity conservation will enable a return to natural conditions, however, or that the "restoration" of indigenous cultures and languages can resurrect them as genuinely autonomous social engagements with local environments: there will be no return to Arcady. The sense of human cultures as multiply differentiated belief and language systems in specific adaption to local environments is increasingly a thing of the past. We are now all collectively moving more toward Deleuzean social assemblages (see, e.g., Ong and Collier 2004) and Appadurai's global ethnoscapes (Appadurai 1996).

Preservation will surely be possible if we work hard enough at it, though that work is a good deal harder than parachuting charismatic species into insufficiently considered cross-cutting assemblages of social interests and material praxis. And while there are effective models for recuperating indigenous languages (like the mentor program in Native California, for example [Hinton et al. 2002]), the rapidity of Native language loss, even in large, very active speech communities like Hopi and Navajo, is not encouraging in the face of the technical mechanisms of popular cultural production and dissemination. Those technical mechanisms and assemblages cannot just be dismissed: while embodying the tools of the enemy (to traditional culture), language revitalization programs will need to rely on the new technology and techniques to effectively reseed those languages into the minds of the younger generations: Inuit TV and Hopi Radio are some limited but useful examples.[2]

We should be skeptical, though, of trusting that the United Nations will provide the road to salvation. The UN Declaration on the Rights of Indigenous Peoples, while perhaps a useful gesture, is ultimately something of an oxymoron. It is exactly nations, nation-states, and their unifications which have created both the problems affecting indigenous peoples and (as the authorizing nexuses of political economies) the problems of natural species endangerment. Perhaps if nation-states agreed to abandon their nationhood, abolish their national laws and political economies, we might trust that their pronouncements could be effective. While demurrals from this declaration are to be deplored as a matter of principle, they may in fact represent more realist responses than those of nations which heartily endorse its principles in theory, but are constitutively incapable of putting these avowals into practice. Which of the nations that have signed on to the declaration has actively restored, and ceded genuine sovereign control over, land base and resources to one indigenous people—sovereign control that can be predicted to last past the next energy crisis?

And at the last, while recognizing the vital need for conservations both cultural and biological, we should be reflexively critical of our motives. Nostalgia for the wild—an untrammeled past of valorized differences, natural or cultural—is based on idealist retrojections. Even if we could get it, which past would we really want, and why?

NOTES

1. Most of my work as an ethnographer has been with the Hopi, but I have conducted some ethnographic fieldwork also with the Hupa.

2. Other examples are listed on the website of the Resource Network for Linguistic Diversity ("Indigenous Languages on Television and Radio," http://www.rnld.org/node/147).

REFERENCES

Appadurai, Arjun
1996 Modernity at Large: Cultural Dimensions of Globalization. Minneapolis: University of Minnesota Press.

Beaglehole, Ernest
1936 Hopi Hunting and Hunting Ritual. Yale University Publications in Anthropology 4. New Haven: Yale University Press.

Bradfield, Richard M.
1973 A Natural History of Associations: A Study in the Meaning of Community. 2 vols. London: Duckworth.

Descola, Philippe
1986 In the Society of Nature. New York: Cambridge University Press.

Hinton, Leanne, Matt Vera, and Nancy Steele
2002 How to Keep Your Language Alive: A Commonsense Approach to One-on-One Language Learning. Berkeley: Heyday Books.

Leach, Edmund
2001[1964] Anthropological Aspects of Language: Animal Categories and Verbal Abuse. In The Essential Edmund Leach. Steven Hugh-Jones and James Laidlaw, eds. Vol. 1, pp. 322–343. New Haven: Yale University Press.

Lévi-Strauss, Claude
    1969 Mythologiques, vol. 1: The Raw and the Cooked. John and Doreen
    Weightman, trans. New York: Viking Penguin.

Montaigne, Michel de
    1993 The Complete Essays. M. A. Screech, trans. London: Penguin Books.

Ong, Aihwa, and Stephen J. Collier, eds.
    2004 Global Assemblages: Technology, Politics, and Ethics as Anthropological
    Problems. New York: Wiley-Blackwell.

Smith, Jonathan Z.
    1972 I Am a Parrot (Red). History of Religions 11(4):391–413.

United Nations Development Programme
    2010 Equator Prize Winners Honoured: 25 Equator Prize Winners Honoured
    for Saving the Environment and Reducing Poverty. September 20. http://con-
    tent.undp.org/go/newsroom/2010/september/25-equator-prize-winners-for-
    the-environment-and-poverty-reduction.en, accessed November 14, 2010.

Vizenor, Gerald
    1999 Manifest Manners: Narratives on Postindian Survivance. Lincoln: Uni-
    versity of Nebraska Press.

Whiteley, Peter M.
    2003 Do "Language Rights" Serve Indigenous Interests? Some Hopi and Other
    Queries. American Anthropologist 105(4):712–722.
    2004 Why Anthropology Needs More History. Journal of Anthropological
    Research 60(4):487–514.

Willis, Roy
    1974 Man and Beast. New York: Basic Books.

CONTRIBUTORS

Janet Chernela is Professor of Anthropology and Latin American Studies at the University of Maryland. She has conducted research in the Brazilian Amazon since 1978, focusing on questions of indigenous peoples' perceptions and use of the forests in which they live as well as the articulations between indigenous rights and conservation. She is author of *A Sense of Space: The Wanano Indians of the Brazilian Amazon*.

Jill Constantino is Lecturer in the Department of Anthropology at Harvard University and the Allston Burr Resident Dean of Cabot House. Her research focuses on social conflicts in the Galápagos Islands, constructions of nature, human/animal relationships, and the interface between humans and technology.

Gregory Forth is Professor of Anthropology at the University of Alberta. He is author of *Beneath the Volcano: Religion, Cosmology and Spirit Classification among the Nage of Eastern Indonesia; Dualism and Hierarchy: Processes of Binary Combination in Keo Society; Nage Birds: Classification and Symbolism among an Eastern Indonesian People;* and *Images of the Wildman in Southeast Asia: An Anthropological Perspective*.

Paul B. Garrett is Associate Professor and Director of Graduate Studies in the Department of Anthropology at Temple University. He is a linguistic anthropologist whose interests include creolization and other language contact phenomena, language socialization, ideologies of language, the political economy of language, and the Caribbean region.

Laurie R. Godfrey is Professor of Anthropology at the University of Massachusetts, Amherst. She is a biological anthropologist and paleontologist with interests in nonhuman primate anatomy and evolution. Her books include *Scientists Confront Creationism* and *What Darwin Began: Modern Darwinian and Non-Darwinian Perspectives on Evolution.*

Michael Hathaway is Assistant Professor of Anthropology at Simon Fraser University. His research explores the politics of nature and indigeneity in China as well as the transnational flows of commercial goods, environmental beliefs, and scientific practices.

Tracey Heatherington is Associate Professor of Anthropology at the University of Wisconsin, Milwaukee and author of *Wild Sardinia: Indigeneity and the Global Dreamtimes of Environmentalism.* Her scholarship considers the cultural politics and postnational contexts of biodiversity conservation, engaging a humanistic approach to the field of sustainable development.

Bernard C. Perley is Associate Professor of Anthropology at the University of Wisconsin, Milwaukee. His research interests include intertextuality, intermediality, and indigeneity as practices of Native American linguistic and cultural revitalization and self-determination. He is Wəlastəkwi (Maliseet) and a member of Tobique First Nation.

Emilienne Rasoazanabary received her Ph.D. from the University of Massachusetts, Amherst. She has studied mouse lemurs since 1999 and has also focused on human activities and attitudes toward conservation, and how human behavior influences the behavior and demography of mouse lemurs.

Genese Marie Sodikoff is Assistant Professor in the Department of Sociology and Anthropology at Rutgers University, Newark. Since 1994, she has studied rain forest conservation and international development, biodiversity conservation, colonial labor regimes, land ethics, and human/animal relations in Madagascar. She is author of *Forest and Labor in Madagascar: From Colonial Concession to Global Biosphere* (IUP, 2012).

Peter M. Whiteley is Curator of North American Ethnology at the American Museum of Natural History in New York City. His principal research addresses society, culture, environmental relations, and history among the Hopi Indians of northern Arizona, where he has conducted extensive ethnographic fieldwork since 1980. Major works include *Rethinking Hopi Ethnography* and *The Orayvi Split: A Hopi Transformation.*

INDEX

Milton Keynes UK
Ingram Content Group UK Ltd.
UKHW022217061223
433909UK00009B/500